BestMasters

Mit „BestMasters" zeichnet Springer die besten Masterarbeiten aus, die an renommierten Hochschulen in Deutschland, Österreich und der Schweiz entstanden sind. Die mit Höchstnote ausgezeichneten Arbeiten wurden durch Gutachter zur Veröffentlichung empfohlen und behandeln aktuelle Themen aus unterschiedlichen Fachgebieten der Naturwissenschaften, Psychologie, Technik und Wirtschaftswissenschaften. Die Reihe wendet sich an Praktiker und Wissenschaftler gleichermaßen und soll insbesondere auch Nachwuchswissenschaftlern Orientierung geben.

Springer awards "BestMasters" to the best master's theses which have been completed at renowned Universities in Germany, Austria, and Switzerland. The studies received highest marks and were recommended for publication by supervisors. They address current issues from various fields of research in natural sciences, psychology, technology, and economics. The series addresses practitioners as well as scientists and, in particular, offers guidance for early stage researchers.

More information about this series at http://www.springer.com/series/13198

Christian Lindorfer

The Language of Self-Avoiding Walks

Connective Constants of Quasi-Transitive Graphs

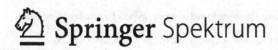
Springer Spektrum

Christian Lindorfer
Graz, Austria

ISSN 2625-3577   ISSN 2625-3615   (electronic)
BestMasters
ISBN 978-3-658-24763-8   ISBN 978-3-658-24764-5   (eBook)
https://doi.org/10.1007/978-3-658-24764-5

Library of Congress Control Number: 2018963840

Springer Spektrum

This Springer Spektrum imprint is published by the registered company Springer Fachmedien Wiesbaden
GmbH part of Springer Nature
The registered company address is: Abraham-Lincoln-Str. 46, 65189 Wiesbaden, Germany

Acknowledgements

I would first and foremost like to thank my thesis advisor Univ.-Prof. Dipl.-Ing. Dr.rer.nat Wolfgang Woess of the Institute for Discrete Mathematics at TU Graz. He allowed this thesis to be my own work, but helped me out whenever I had problems or just needed someone to discuss the content of this thesis.

To my very good friend and fellow student Stefan Hammer, thank you for listening and offering me advice during the whole writing process and also for pointing out some mistakes in this document. I am also very grateful to my close friend Daniel Hischenhuber for the constructive criticism of the manuscript he provided.

Lastly, I would like to thank my parents for all the moral and financial support and the amazing chances they have given me over the years.

Christian Lindorfer

Contents

List of Figures

List of Tables

1. Introduction

Imagine the following discrete process: You are starting at a vertex of a given undirected graph. In every step you can choose any edge leaving the current vertex and follow this edge to a new vertex. The only rule is that you must not return to any vertex already visited during the process. We call such walks self-avoiding. This process leads to the following question:

For a given number n, how many possible paths of length n could you have followed?

The answer to this question for all n is only known for some special graphs, the general case seems very difficult to solve. A less difficult question would be to ask for the asymptotic behaviour of the number of paths for n going to infinity. Clearly this is only interesting for graphs having infinitely many vertices. Although this new question is easier to answer than the original one, it still provides a hard task. The most important graphs in this theory are the integer lattices $\mathbb{Z}^d$ (especially $d = 2$ and $d = 3$). A lot of work has been dedicated to getting the asymptotic growth rate of the number of self-avoiding walks, but still much remains unknown. Many interesting results on this topic can be found in [19].

Self-avoiding walks were introduced in 1953 as a model for long-chain polymer molecules by the famous chemist Paul J. Flory in [9]. Polymer scientists want to know how many different configurations a polymer chain consisting of n monomers can have. Although polymer chains live in the continuum, in many cases a lattice approximation is good enough. The self avoidance models the excluded volume effect: No two monomers can be at the same position. Since then, self-avoiding walks have become very important in statistical physics, for example in percolation theory. More about percolation can be found in [10].

In this thesis we start by introducing the notions of self-avoiding walks and the connective constant of quasi-transitive graphs. We will prove the existence of the connective constant following a result of Hammersley [16] from 1957. We give some examples of graphs where the connective constant or good bounds for it are already known and also some general bounds holding for graphs of certain types. One of the most important results in this topic was the paper [7] of Duminil-Copin and Smirnov, which was published in 2010 and contained the first rigorous proof for the connective constant of the honeycomb lattice being $\sqrt{2 + \sqrt{2}}$. In this thesis we use

© Springer Fachmedien Wiesbaden GmbH, part of Springer Nature 2018
C. Lindorfer, *The Language of Self-Avoiding Walks*, BestMasters,
https://doi.org/10.1007/978-3-658-24764-5_1

this result and some theory about generating functions to calculate the connective constant of the Archimedian lattice $(3, 12^2)$. This result without the detailed proof was given by Grimmett in [11].

Another important concept when working with self-avoiding walks are bridges. Bridges are a subclass of all self-avoiding walks and under certain circumstances it can be shown that the bridge constant of a graph, which is defined similarly to the connective constant, equals the connective constant. The statement is called Bridge Theorem and was proved by Grimmett and Li in [13]. We use it to calculate the connective constants of the integer strips $\mathbb{Z} \times \{0, 1\}$ and $\mathbb{Z} \times \{0, 1, 2\}$, which are sub-lattices of the integer lattice $\mathbb{Z}^2$. We also prove, mostly following Beffara and Huynh in [2], that the connective constants of these integer strips $\mathbb{Z} \times \{0, 1, \ldots n\}$ converge to the connective constant of the integer lattice $\mathbb{Z}^2$, when sending n to infinity.

In Chapter 4 we follow the work of Alm and Janson in [1] to prove that for any one-dimensional lattice, the connective constant is an algebraic number. It is still an open problem whether this is also true for the integer lattice $\mathbb{Z}^2$.

Our goal in the next part of this thesis is to use context-free languages to describe the set of all self-avoiding walks on a given graph. We start in Chapter 5 by introducing different types of grammars and their generated languages. We also give an introduction into the theory of generating functions of context-free languages, which was developed by Chomsky and Schützenberger in [4].

In Chapter 6 we start by defining "good labellings" of graphs and what we mean when talking about the language of self-avoiding walks. We will then use the theory about context free languages to get unambiguous grammars for the language of bridges and the language of self-avoiding walks on the ladder graph produced by these grammars and their corresponding generating functions. Finally we consider the resulting graph when taking two copies of the infinite k-regular tree $\mathbb{T}_k$ for arbitrary k and connecting every pair of vertices corresponding to the same vertex in the original tree. Again we give an unambiguous grammar generating the language of self-avoiding walks and solve the resulting system of equations to get the generating function and thereby the connective constant depending on k. This idea of using language theory and grammars to get generating functions for the number of self-avoiding walks is quite new and has not been studied so far. Hopefully it can be used to get some new interesting results.

2. Self-avoiding walks and connective constants

2.1. Existence on quasi-transitive graphs

Definition 1. A graph $G = (V, E)$ consists of a finite or countably infinite set of vertices V and a set of edges $E \subset V \times V$ connecting the vertices. We call G *simple*, if there are no loops, i.e. no edges of the form (v, v) for $v \in V$, in G. For an edge $e = (u, v)$ we denote by $e^- = u$ its starting point and by $e^+ = v$ its endpoint. We call G *undirected* if for all $v, w \in V$

$$(v, w) \in E \text{ if and only if } (w, v) \in E.$$

In this case we will denote the undirected edge corresponding to the pair (v, w), (w, v) by $\{v, w\}$.

A *walk* on $G = (V, E)$ is a sequence $\pi = (v_0, v_1, \ldots, v_n)$ with $v_i \in V$ for $0 \leq i \leq n$ and $(v_{i-1}, v_i) \in E$ for $1 \leq i \leq n$. The length of the walk π is denoted by $|\pi| = n$ and we call π an n-step walk connecting v_0 and v_n. We say that G is *connected* if for all $v, w \in V$ there is a walk connecting v and w in G. The distance $d(v, w)$ of v and w is equal to k, if the shortest walk connecting v and w has length k.

Let now G be an undirected graph. For any vertex $v \in V$ and edge $e \in E$ we say that v and e are *incident*, if $e = \{v, w\}$ for some $w \in V$. We denote the number of $w \in V$ with $\{v, w\} \in E$ by $deg(v)$ and call it the *degree* of v. G is said to be *locally finite*, if $deg(v) < \infty$ for each $v \in V$ and *k-regular*, if $deg(v) = k$ for all $v \in V$.

If not mentioned otherwise the graphs used here are simple, undirected, locally finite and connected. As already mentioned we are mostly interested in graphs with infinitely many vertices. To make sure that the connective constant exists on graphs considered in this thesis, we want them to be quasi-transitive as defined next.

Definition 2. The *automorphism group* of a graph $G = (V, E)$, denoted by $AUT(G)$, is the group of all permutations $\sigma : V \to V$ such that for all $u, v \in V$ we have: $\{u, v\} \in E$ if and only if $\{\sigma(u), \sigma(v)\} \in E$.

A subgroup $\Gamma \leq AUT(G)$ is said to act *transitively* on G if, for any $u, v \in V$, there exists $\gamma \in \Gamma$ with $\gamma u = v$. It is said to act *quasi-transitively* if there exists a finite set $W \subset V$ such that for any $u \in V$ there exist $v \in W$

© Springer Fachmedien Wiesbaden GmbH, part of Springer Nature 2018
C. Lindorfer, *The Language of Self-Avoiding Walks*, BestMasters,
https://doi.org/10.1007/978-3-658-24764-5_2

and $\gamma \in \Gamma$ with $\gamma u = v$.

A graph G is called *transitive* (respectively *quasi-transitive*) if $AUT(G)$ acts transitively (respectively quasi-transitively) on G.

The Γ-*stabilizer* $Stab_v^\Gamma$ of $v \in V$ is the set of $\gamma \in \Gamma$ for which $\gamma v = v$. The orbit Γv of $v \in V$ under the action of Γ is defined as the set of all γv for $\gamma \in \Gamma$.

Remark 1. For a given graph $G = (V, E)$ we can define an equivalence relation $\sim$ on V by $u \sim v$ if and only if there is a $\gamma \in AUT(G)$ such that $\gamma u = v$. By definition the orbit $AUT(G)v$ of a vertex $v \in V$ under the action of $AUT(G)$ is an equivalence class of this relation. If there are only finitely many orbits, we can choose any set of representatives of the equivalence classes as the set W in the definition of quasi-transitivity. Therefore G is quasi-transitive if and only if the number of orbits is finite. For transitive graphs there is exactly one orbit.

Definition 3. A walk on a graph $G = (V, E)$ is called *self-avoiding (SAW)* if it visits no vertex more than once.

We denote by $\Sigma_n(v)$ the set of SAWs of length $n \geq 0$ on G starting at the vertex $v \in V$ and by $\sigma_n(v) = |\Sigma_n(v)|$ its cardinality.

For graphs of interest, the number of self-avoiding walks $\sigma_n(v)$ grows exponentially fast for every $v \in V$. First we want to show that the limit of $\sigma_n(v)^{1/n}$ for n going to infinity exists and that it is independent of the choice of v under the condition of G being quasi-transitive. For this we need Fekete's Lemma about the limit of subadditive sequences.

Lemma 1. Let $(a_n)_{n \geq 1}$ be a sequence of real numbers which is subadditive, i.e., $a_{n+m} \leq a_n + a_m$ for all integers $n, m \geq 1$. Then the limit $\lim_{n \to \infty} n^{-1} a_n$ exists in $[-\infty, \infty)$ and we get

$$\lim_{n \to \infty} \frac{a_n}{n} = \inf_{n \geq 1} \frac{a_n}{n}. \tag{2.1}$$

Proof. It suffices to show that

$$\limsup_{n \to \infty} \frac{a_n}{n} \leq \frac{a_k}{k} \quad \text{for every integer } k \geq 1 \tag{2.2}$$

since we get the existence of the limit by taking the $\liminf_{k \to \infty}$ in (2.2) and then (2.1) can be seen by taking the $\inf_{k \geq 1}$ in (2.2).

For showing (2.2), we fix some $k \geq 1$ and let

$$A_k := \max_{1 \leq r \leq k} a_r.$$

For a given integer $n \geq 1$ let the integers $q \geq 0$ and $r \in \{1, \ldots, k\}$ be such that $n = qk + r$. By subadditivity we have

$$a_n \leq qa_k + a_r \leq \frac{n}{k}a_k + A_k.$$

Dividing by n and taking the $\limsup_{n\to\infty}$ proves (2.2):

$$\limsup_{n\to\infty} \frac{a_n}{n} \leq \frac{a_k}{k} + \lim_{n\to\infty} \frac{A_k}{n} = \frac{a_k}{k}.$$

$\square$

Using this result it is not difficult to prove the existence of the connective constant for transitive graphs. Here we want to have the following more general result proved by Hammersley [16] in 1957, which shows the existence of the connective constant for all quasi-transitive graphs.

Theorem 1. *Let $G = (V, E)$ be an infinite quasi-transitive graph. Then there exists $\mu = \mu(G) \in [1, \infty)$, called the connective constant of G, such that*

$$\mu = \lim_{n\to\infty} \sigma_n(v)^{\frac{1}{n}} \quad \text{for all } v \in V.$$

Proof. The action of $AUT(G)$ on the graph G admits finitely many orbits $\Gamma_1, \Gamma_2, \ldots, \Gamma_N$ because G is quasi-transitive. Let $\{v_1, v_2, \ldots, v_N\}$ with $v_i \in \Gamma_i$ be a set of representatives of the N orbits. We define

$$\sigma_n := \max_{1 \leq i \leq N} \sigma_n(v_i) \quad \text{for all integers } n \geq 0 \tag{2.3}$$

and note that σ_1 is the max degree of G.

Our first goal is to show that $\sigma_n(v) \geq 1$ for all $v \in V$ and integers $n \geq 0$. Let $B_n(v) = \{w \in V \mid d(v, w) \leq n\}$ be the ball of radius n centered in v. Then $1 \leq |B_n(v)| \leq \sigma_1^n + 1$, where the right inequality holds because σ_1 is the max degree in G. Since G is infinite and connected, there are $x \in B_n(v)$ and $y \in V \setminus B_n(v)$ such that $\{x, y\} \in E$. There is a walk π of length $\leq n$ connecting v and x in G. Then $|\pi| = n$ as otherwise we would get $y \in B_n(v)$. Also π is a SAW, because if π visits a vertex twice we can remove the cycle and get a shorter walk connecting v and x. Using the fact that v is in Γ_i for some $1 \leq i \leq N$ and therefore $\sigma_n(v) = \sigma_n(v_i)$ we get

$$1 \leq \sigma_n(v) \leq \sigma_n \quad \text{for all } v \in V, \ n \geq 0. \tag{2.4}$$

Now for given $v \in V$ and integers $n, m \geq 0$ each $(m+n)$-step SAW starting at v can be seen as a concatenation of an m-step SAW starting at v and ending at some $w \in V$ and an n-step SAW starting at w. We obtain the estimates

$$\sigma_{n+m}(v) \leq \sigma_m(v)\sigma_n \leq \sigma_m\sigma_n \quad \text{for all } v \in V, \ m, n \geq 0. \qquad (2.5)$$

This naturally holds for the v maximizing the left hand side, so it follows that

$$\sigma_{m+n} \leq \sigma_m\sigma_n \quad \text{for all } m, n \geq 0,$$

which is equivalent to $\log \sigma_n$ being a subadditive sequence. Using Lemma 1 and (2.4) we get

$$\lim_{n \to \infty} \frac{\log \sigma_n}{n} = \inf_{n \geq 1} \frac{\log \sigma_n}{n} \geq 0.$$

We can define the connective constant μ as the exponential of the above limit:

$$\mu := \lim_{n \to \infty} \sigma_n^{\frac{1}{n}} \geq 1.$$

For every $\lambda > \mu$ there exists a constant $C = C(\lambda) \geq \lambda \geq 1$ such that

$$\sigma_n(v) \leq \sigma_n \leq C\lambda^n \quad \text{for all } v \in V, \ n \geq 0. \qquad (2.6)$$

The inequalities (2.5) and (2.6) imply

$$\sigma_{n+m}(v) \leq C\lambda^n\sigma_m(v) \quad \text{for all } v \in V, \ m, n \geq 0. \qquad (2.7)$$

Let $u, v \in V$ with $e = \{u, v\} \in E$. Let π be a $2n$-step SAW starting in u. We can distinguish two cases:

1. If π does not meet v, we add e in front of π and get a $(2n + 1)$-step SAW starting at v.

2. If π meets v after $k \leq 2n$ steps, we can view the first part of π as a k-step SAW connecting v and u and the second part as a $(2n - k)$-step SAW starting in v (possibly with length 0).

It follows that

$$\sigma_{2n}(u) \leq \sigma_{2n+1}(v) + \sum_{k=1}^{2n} \sigma_k(v)\sigma_{2n-k}(v). \qquad (2.8)$$

Application of (2.6) and (2.7) on the right hand side of (2.8) gives

$$\sigma_{2n}(u) \leq C\lambda^{n+1}\sigma_n(v) + \sum_{k=1}^{n} C\lambda^k C\lambda^{n-k}\sigma_n(v) + \sum_{k=n+1}^{2n} C\lambda^{k-n}\sigma_n(v)C\lambda^{2n-k}$$

$$\leq (2n+1)C^2\lambda^n\sigma_n(v) \quad \text{for all } \{u, v\} \in E, \ n \geq 0.$$

$$(2.9)$$

Let $u, v \in V$ be two vertices with $d(u,v) = d$. Consecutively using (2.9) along a walk of length d connecting u and v in G yields

$$\sigma_{2^d n}(u) \le \sigma_n(v) \prod_{i=1}^{d} [(2^i n + 1)C^2 \lambda^{2^{i-1}n}] \le \sigma_n(v) \prod_{i=1}^{d} [3^i n C^2 \lambda^{2^{i-1}n}] \tag{2.10}$$
$$= 3^{d(d+1)/2} C^{2d} n^d \lambda^{(2^d-1)n} \sigma_n(v) \quad \text{for all } n \ge 0.$$

Fix a vertex $v \in V$ and let $D = D(v) := \max\limits_{1 \le i \le N} d(v, v_i)$. Then for $u \in V$ with $d(u,v) = d \le D$ it follows from (2.7) and (2.10) that

$$\sigma_{2^D n}(u) \le C \lambda^{(2^D - 2^d)n} \sigma_{2^d n}(u) \le \widehat{C} n^D \lambda^{(2^D-1)n} \sigma_n(v),$$

where $\widehat{C}$ is a constant depending only on $D(v)$ and $C(\lambda)$. By 2.3 and the definition of D, in particular

$$\sigma_{2^D n} = \max_{1 \le i \le N} \sigma_{2^D n}(v_i) \le \widehat{C} n^D \lambda^{(2^D-1)n} \sigma_n(v). \tag{2.11}$$

Using the definition of μ and (2.11) we get

$$\log \mu = \lim_{n \to \infty} \frac{1}{2^D n} \log \sigma_{2^D n}$$
$$\le \liminf_{n \to \infty} \frac{1}{2^D n} [\log \widehat{C} + D \log n + (2^D - 1)n \log \lambda + \log \sigma_n(v)]$$
$$\le \log \lambda + \frac{1}{2^D} [-\log \lambda + \liminf_{n \to \infty} \frac{1}{n} \log \sigma_n(v)].$$

Because D is independent of λ, sending $\lambda \to \mu$ yields

$$\log \mu \le \liminf_{n \to \infty} \frac{1}{n} \log \sigma_n(v)$$

and therefore also

$$\mu \le \liminf_{n \to \infty} \sigma_n(v)^{\frac{1}{n}}. \tag{2.12}$$

But we already know from (2.4) that

$$\limsup_{n \to \infty} \sigma_n(v)^{\frac{1}{n}} \le \lim_{n \to \infty} \sigma_n^{\frac{1}{n}} = \mu. \tag{2.13}$$

By (2.12) and (2.13), $\lim\limits_{n \to \infty} \sigma_n(v)^{\frac{1}{n}}$ exists and is equal to μ. $\qquad\square$

We can start to calculate the connective constant for some simple graphs by counting SAWs starting at some fixed vertex v. By Theorem 1 the choice of v does not change the result.

Example 1. For two integers $k, l \geq 2$ the bi-regular tree $\mathbb{T}_{k,l}$ is an infinite tree where the vertex degree is constant on each of the two bipartite classes, with values k and l, respectively. We count the number of SAWs of length n starting at a given vertex of degree k. For the first step, we have k possibilities, for all subsequent steps alternately $l - 1$ and $k - 1$, because we can never go back and visit a vertex again. This gives

$$\sigma_n(v) = k(k-1)^{\lfloor \frac{n-1}{2} \rfloor} (l-1)^{\lceil \frac{n-1}{2} \rceil}.$$

We can now calculate the connective constant:

$$\mu(\mathbb{T}_{k,l}) = \lim_{n \to \infty} \sigma_n(v)^{\frac{1}{n}} = \sqrt{(k-1)(l-1)}.$$

Remark 2. Quasi-transitivity plays an important role in this theory. The following example shows that there are (non-quasi-transitive) graphs for which the connective constant does not exist in $[1, \infty)$.

Let $\mathbb{T}$ be a rooted tree with root v, where v has degree one and every vertex u at distance $d > 0$ from v has degree $d + 2$ as shown in Figure 2.1.

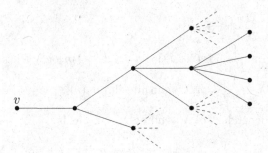

Figure 2.1: Infinite tree $\mathbb{T}$: The number of children increases by 1 in every step.

Obviously $\mathbb{T}$ is not quasi-transitive as it contains vertices with arbitrary big degrees and therefore infinitely many orbits. By counting SAWs as in Example 1 we get

$$\lim_{n \to \infty} (\sigma_n(v))^{1/n} = \lim_{n \to \infty} \left(\prod_{i=1}^{n} i \right)^{1/n} = \infty.$$

2.2. Some known values and simple bounds

There are a number of graphs for which the connective constant is already known. The most famous ones are the ladder $\mathbb{L}$ and the honeycomb lattice $\mathbb{H}$ (see Figure 2.2), for which

$$\mu(\mathbb{L}) = \frac{1}{2}(1 + \sqrt{5}), \quad \mu(\mathbb{H}) = \sqrt{2 + \sqrt{2}}.$$

For the ladder, it is not very difficult to count the number of SAWs directly (for example by using generating functions [22]). On the other hand, the connective constant of the honeycomb lattice is a lot harder to get and the first mathematical proof was provided in [7] by Duminil-Copin and Smirnov in 2010 using the so-called parafermionic observable and bridges, which will be introduced in Chapter 3.

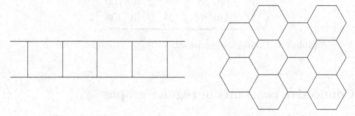

Figure 2.2: Ladder graph $\mathbb{L}$ and honeycomb lattice $\mathbb{H}$.

In contrast, the connective constants of the square lattice $\mathbb{Z}^2$ and also the higher dimensional integer lattices $\mathbb{Z}^d$ for $d \geq 2$ are still unknown and a lot of work has been dedicated to finding good bounds for them. A nice general discussion about self-avoiding walks on integer lattices can be found in the book [19] by Madras and Slade. Some currently known good bounds for the connective constant of the two dimensional case are

$$2.6256 \leq \mu(\mathbb{Z}^2) \leq 2.6792.$$

The lower bound was given by Jensen [18] by enumeration of bridges. Pönitz and Tittmann [21] proved the upper bound using finite automata to construct SAWs with finite memory, which are walks where vertices can reappear after a given number of steps.

Remark 3. We see that the connective constants of the ladder and the honeycomb lattice are algebraic numbers. For the square lattice this question is still open. Using numerical estimations from the 1980s it was believed for about 30 years, that $\mu(\mathbb{Z}^2)$ could be a root of the polynomial

$13x^4 - 7x^2 - 581$. This however does not seem to be true using current good estimates, for example the one in [17] by Jacobsen, Scullard and Guttmann, which differ from the predicted value in the twelfth digit.

Of great interest are also the integer strips $SQ_{0,n} = \mathbb{Z} \times \{0, 1, \ldots, n\}$ for $n \geq 1$. In Section 3 we will show that the sequence of connective constants $\mu(SQ_{0,n})$ converges to $\mu(\mathbb{Z}^2)$ for n going to infinity. Table 2.1 is taken from [1]; parts of the method Alm and Janson used to find the values will be discussed in Chapter 4.

n	$\mu(SQ_{0,n})$	n	$\mu(SQ_{0,n})$
0	1	5	2.276379
1	1.618034	6	2.332779
2	1.914627	7	2.375451
3	2.087285	8	2.408709
4	2.198966	9	2.435258

Table 2.1: Connective constants of $SQ_{0,n}$ (rounded values).

2.3. Connective constants of regular graphs

Next we want to give bounds for the connective constant of regular graphs. It is easy to see that for any infinite quasi-transitive Δ-regular graph G we have

$$1 \leq \mu(G) \leq \Delta - 1. \tag{2.14}$$

The second inequality is obtained by the following upper bound on the number of SAWs of length n in G as in Example 1:

$$\sigma_n(v) \leq \Delta(\Delta - 1)^{n-1}.$$

Both of these bounds are already tight for quasi-transitive graphs. For the upper bound we can just consider the regular tree $\mathbb{T}_\Delta$ of degree Δ. As in Example 1 we get $\mu(\mathbb{T}_\Delta) = \Delta - 1$. For showing the tightness of the lower bound we consider a graph where an infinite line is decorated with finite graphs attached by exactly one vertex as shown in Figure 2.3. This is possible for all degrees $\Delta \geq 3$.

Let the attached graphs have k vertices. They are attached to the line at exactly one vertex, so SAWs of length n starting at some vertex v of the line cannot leave the line and come back to it, otherwise a vertex would

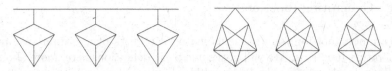

Figure 2.3: Infinite quasi-transitive regular graphs of degree 3 and 4, following [14].

appear twice. So for large n, the SAWs need to follow the line (in one of the two directions) and can maybe have an end-piece of length $\leq k$ in one of the attached graphs. Therefore we have

$$\mu = \lim_{n\to\infty} \sigma_n(v)^{1/n} \leq \lim_{n\to\infty} (2(\Delta - 1)^k)^{1/n} = 1.$$

The upper bound in (2.14) is only achieved by the Δ-regular tree. This follows from the following theorem proved by Grimmett and Li in [14].

Theorem 2. *Let $G = (V, E)$ be an infinite, quasi-transitive graph and let $\Delta \geq 3$. Then $\mu(G) < \Delta - 1$ if one of the following conditions holds:*

(a) G is Δ-regular and contains a cycle,

(b) $deg(v) \leq \Delta$ for all $v \in V$, and there exists a vertex $u \in V$ with $deg(u) \leq \Delta - 1$.

Improved lower bounds may be achieved when considering only transitive graphs. Grimmett and Li showed in [14] the following theorem:

Theorem 3. *Let for an integer $\Delta \geq 2$ and G be an infinite Δ-regular transitive graph. Then*

$$\mu(G) \geq \sqrt{\Delta - 1}.$$

The only value of Δ for which the lower bound is known to be tight is $\Delta = 2$. For $\Delta = 3$ there is evidence for the improved lower bound

$$\mu(G) \geq \frac{1}{2}(1 + \sqrt{5})$$

being true for all transitive 3-regular graphs G. It has been shown in [15] that this inequality must hold for some specific types of graphs, but the general question is still open. Moreover we already know that this value is achieved by the ladder graph.

2.4. Generating functions

A common tool for counting combinatorial objects of a certain size are generating functions. We will give some basic definitions here; more about analytic combinatorics can be found in [8].

Definition 4. A *combinatorial class* $(A, |\cdot|)$ is a finite or countably infinite set A on which a size function $|\cdot|$ satisfying the following conditions is defined:

(a) The size of an element is a non-negative integer.

(b) The number of elements of any given size is finite.

We denote the size of an element $\alpha \in A$ by $|\alpha|$, by A_n the subset of elements of A having size n and by a_n its cardinality $|A_n|$.
The *ordinary generating function* $F_A(z)$ of the combinatorial class A is the formal power series

$$F_A(z) = \sum_{n=0}^{\infty} a_n z^n = \sum_{a \in A} z^{|a|}.$$

Remark 4. Let two combinatorial classes $(A, |\cdot|_A)$ and $(B, |\cdot|_B)$ and their generating functions $F_A(z)$ and $F_B(z)$ be given.
The combinatorial sum of the classes A and B is the class $(S, |\cdot|_S)$, where $S = A \cup B$ is a disjoint union and for $\omega \in A$

$$|\omega|_S = \begin{cases} |\omega|_A & \text{if } \omega \in A, \\ |\omega|_B & \text{if } \omega \in B. \end{cases}$$

Then we get the ordinary generating function $F_S(z)$ of S as the sum of the generating functions of A and B,

$$F_S(z) = F_A(z) + F_B(z).$$

The cartesian product of the classes A and B is the class $(P, |\cdot|_P)$ where $P = A \times B$ and for $\omega = (\alpha, \beta) \in P$

$$|\omega|_P = |\alpha|_A + |\beta|_B.$$

Then the ordinary generating function $F_P(z)$ of P is the product of the generating functions of A and B,

$$F_P(z) = F_A(z) \cdot F_B(z).$$

Given the class A not containing the element ϵ of size 0 we define $A^* = \bigcup_{k \geq 0} A^k$, where A^0 denotes the class containing only ϵ and A^k denotes the k-fold cartesian product $A \times \cdots \times A$ for $k \geq 1$. In other words, we have

$$A^* = \{(\alpha_1, \ldots, \alpha_k) \mid k \geq 0, \alpha_i \in A \text{ for } 1 \leq i \leq k\}.$$

The sequence class of A is $(A^*, |\cdot|_{A^*})$, where

$$|(\alpha_1, \ldots, \alpha_k)|_{A^*} = |\alpha_1|_A + \cdots + |\alpha_k|_A.$$

For the ordinary generating function F_{A^*} of A^* we get that

$$F_{A^*}(z) = 1 + F_A(z) + (F_A(z))^2 + (F_A(z))^3 + \cdots = \frac{1}{1 - F_A(z)}.$$

Clearly for a given graph $G = (V, E)$ and a vertex $v \in V$, the set of self-avoiding walks $\Sigma(v)$ starting at v together with the size function $|\cdot|$ mapping a path to its length is a combinatorial class. We introduce its ordinary generating function

$$F_{\Sigma(v)}(z) = \sum_{n=0}^{\infty} \sigma_n(v) z^n.$$

Remark 5. Using the existence of the limit and the definition of the connective constant $\mu(G)$ in Theorem 1 and the Cauchy Hadamard formula we get that the power series $F(t)$ has radius of convergence

$$R = \frac{1}{\lim\limits_{n \to \infty} |\sigma_n(v)|^{1/n}} = \frac{1}{\mu(G)}.$$

Therefore $F_{\Sigma(v)}(z)$ defines an analytic function in the complex parameter z if $|z| < 1/\mu$. When considering $F_{\Sigma(v)}(x)$ for $x \in \mathbb{R}^+$ we get

$$F_{\Sigma(v)}(x) \begin{cases} < \infty & \text{if } x < 1/\mu, \\ = \infty & \text{if } x > 1/\mu. \end{cases} \qquad (2.15)$$

This fact can be used to check that a given value in $\mathbb{R}^+$ is the connective constant of a graph. This idea was used by Duminil-Copin and Smirnov in their proof that the connective constant of the honeycomb lattice equals $\sqrt{2 + \sqrt{2}}$, which was already predicted by Nienhuis in [20].

We also know that our generating functions have only positive coefficients. So the following theorem known as Pringsheim's theorem can be applied yielding a singularity at $z = R = 1/\mu(G)$.

Theorem 4. *If the complex function $f(z)$ is representable at the origin by a series expansion that has non-negative coefficients and radius of convergence R, then the point $z = R$ is a singularity of $f(z)$.*

The result of the following example without the detailed proof was given by Grimmett in [11]. Now we give a detailed proof of the below statement to give an example of how generating functions can be used to verify the value of the connective constant of a given graph.

Example 2. The Archimedian lattice $(3, 12^2)$ here denoted by $\mathbb{A}$ is obtained by replacing each vertex of the honeycomb lattice $\mathbb{H}$ by a triangle as shown in Figure 2.4. This process is often called Fisher Transformation.

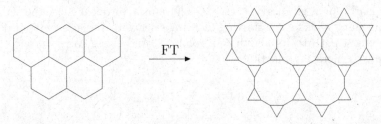

Figure 2.4: Fisher Transformation (FT) of $\mathbb{H}$ yields Archimedian lattice $(3, 12^2)$.

We will now show that $\mu(\mathbb{A})$ satisfies the equation

$$\frac{1}{\mu(\mathbb{A})^2} + \frac{1}{\mu(\mathbb{A})^3} = \frac{1}{\mu(\mathbb{H})}. \tag{2.16}$$

This equation has a unique positive solution for $\mu(\mathbb{A})$ as the polynomial $x^2 + x^3$ is bijective when seen as a function from $\mathbb{R}^+$ to $\mathbb{R}^+$. Grimmett and Li proved in [12] that this equation holds for every infinite quasi-transitive connected 3-regular graph G and its Fisher Transformation $F(G)$.

Edges of $\mathbb{A}$ lying in a triangle are called triangular. For a given $v \in V(\mathbb{H})$ let $\Sigma_\mathbb{H}$ be the set of all SAWs starting at v and for a fixed vertex $w \in V(\mathbb{A})$ in the triangle corresponding to v let $\Sigma_\mathbb{A}$ be the set of all SAWs starting at w. Then

$$F_\mathbb{H}(x) = \sum_{\pi \in \Sigma_\mathbb{H}} x^{|\pi|} \quad \text{and} \quad F_\mathbb{A}(x) = \sum_{\pi \in \Sigma_\mathbb{A}} x^{|\pi|}$$

are the generating functions of SAWs in $\mathbb{H}$ and $\mathbb{A}$ respectively. Let $\Sigma_\mathbb{A}^* \subset \Sigma_\mathbb{A}$ be the subset of SAWs of length ≥ 1 starting and ending at a triangular edge and $F_{\mathbb{A}}^*(x)$ its generating function. As $\Sigma_{\mathbb{A}}^* \subset \Sigma_{\mathbb{A}}$ clearly

$$F_{\mathbb{A}}^*(x) \leq F_{\mathbb{A}}(x) \quad \text{for all} \quad x \in \mathbb{R}^+. \tag{2.17}$$

Next we want to have an upper estimate for the generating function $F_{\mathbb{A}}(x)$ in terms of $F_{\mathbb{A}}^*(x)$. To achieve this we shorten paths in $\Sigma_{\mathbb{A}}$ to get paths π in $\Sigma_{\mathbb{A}}^*$ in the following way shown in Figure 2.5 :

(a) If π starts with a triangular edge, leave the beginning, else remove the first edge $\{w, w'\}$ and get a path starting with a triangular edge at w'.

(b) If π ends with a triangular edge, leave it, else remove the last edge to get a path which ends with a triangular edge.

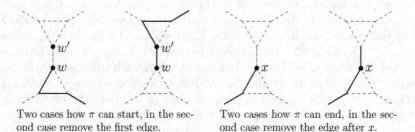

Two cases how π can start, in the second case remove the first edge.   Two cases how π can end, in the second case remove the edge after x.

Figure 2.5: Different cases for initial and final part of paths in $\Sigma_{\mathbb{A}}$.

Observe that the set of SAWs starting at w' with a triangular edge and ending with a triangular edge has the same generating function as $\Sigma_{\mathbb{A}}^*$ since $\mathbb{A}$ is transitive. Therefore all walks in $\Sigma_{\mathbb{A}}$ can be decomposed into an initial part having generating function $1 + x$, a walk in $\Sigma_{\mathbb{A}}^*$ and a final part of length 0 or 1 also having generating function $1 + x$. We get the inequality

$$F_{\mathbb{A}}(x) \leq (1 + x)^2 F_{\mathbb{A}}^*(x) \quad \text{for all } x \in \mathbb{R}^+. \tag{2.18}$$

Now using (2.17) and (2.18) we get for all $x \in \mathbb{R}^+$

$$F_{\mathbb{A}}(x) < \infty \iff F_{\mathbb{A}}^*(x) < \infty$$

and by using (2.15) we conclude

$$F_{\mathbb{A}}^*(x) \begin{cases} < \infty & \text{if } x < 1/\mu(\mathbb{A}), \\ = \infty & \text{if } x > 1/\mu(\mathbb{A}). \end{cases} \tag{2.19}$$

Let $\Sigma_{\mathbb{H}}^* \subset \Sigma_{\mathbb{H}}$ be the set of walks on $\mathbb{H}$ not starting with the edge of $\mathbb{H}$ which is incident to w in $\mathbb{A}$. In $\Sigma_{\mathbb{H}}^*$ exactly one third of all SAWs of length ≥ 1 are missing, because the number of walks starting in each of the three possible directions is equal. The generating function of $\Sigma_{\mathbb{H}}^*$ is

$$F_{\mathbb{H}}^*(x) = 1 + \frac{2}{3}(F_{\mathbb{H}}(x) - 1).$$

We can conclude that the radii of convergence of $F_{\mathbb{H}}^*(x)$ and $F_{\mathbb{H}}(x)$ are equal. By (2.15), we get

$$F_{\mathbb{H}}^*(x) \begin{cases} < \infty & \text{if } x < 1/\mu(\mathbb{H}), \\ = \infty & \text{if } x > 1/\mu(\mathbb{H}). \end{cases} \qquad (2.20)$$

We associate every SAW π^* in $\Sigma_{\mathbb{A}}^*$ with a SAW π in $\Sigma_{\mathbb{H}}^*$ by shrinking all triangles introduced by the Fisher Transformation. Note that this is only possible because π^* starts and ends with a triangular edge and therefore can visit every triangle at most once. Every walk π of length n in $\Sigma_{\mathbb{H}}^*$ arises from 2^{n+2} SAWs in $\Sigma_{\mathbb{A}}^*$. This follows from the fact that any π^* associated to π contains edges of exactly $n+1$ triangles, where each triangle contributes either 1 or 2 edges to π^* and for the last triangle there are altogether 4 possibilities, which edges it can contribute. We do not get any walks in $\Sigma_{\mathbb{H}} \setminus \Sigma_{\mathbb{H}}^*$ as all walks in $\Sigma_{\mathbb{A}}^*$ need to start with a triangular edge incident to w and therefore cannot contain the non-triangular edge incident to w. An example for a path π^* in $\Sigma_{\mathbb{A}}^*$ and its associated path π in $\Sigma_{\mathbb{H}}^*$ is shown in Figure 2.6.

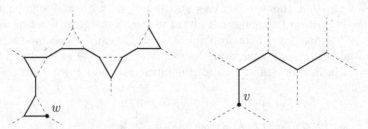

Figure 2.6: A SAW π^* in $\Sigma_{\mathbb{A}}^*$ and its associated SAW π in $\Sigma_{\mathbb{H}}^*$.

This association of SAWs leads to

$$(2x + 2x^2) F_{\mathbb{H}}^*(x(x + x^2)) = F_{\mathbb{A}}^*(x) \qquad (2.21)$$

where the term $(2x + 2x^2)$ in front of the generating function corresponds to the removed edges in the last visited triangle and $(x + x^2)$ corresponds to the lost 1 or 2 edges when shrinking the first n triangles of the walk. Now clearly, the left hand side is $< \infty$, if and only if $F_{\mathbb{H}}^*(x(x + x^2)) < \infty$. Using (2.20) in (2.21) we get

$$F_{\mathbb{A}}^*(x) \begin{cases} < \infty & \text{if } x^2 + x^3 < 1/\mu(\mathbb{H}), \\ = \infty & \text{if } x^2 + x^3 > 1/\mu(\mathbb{H}). \end{cases} \qquad (2.22)$$

Thus the claim (2.16) follows from (2.19) and (2.22). Using this claim we get for the connective constant of the Archimedian lattice $(3, 12^2)$

$$\mu(\mathbb{A}) \approx 1.711041.$$

3. Graph height functions and bridges

3.1. The Bridge Theorem

Counting SAWs in a graph can be rather difficult. We want to reduce the number of walks to be counted by only considering so-called bridges, which form a subclass of all SAWs. The number of bridges of a given length lead to the bridge constant of a graph. Under certain conditions on the graph, its bridge constant is equal to its connective constant. To define bridges and the bridge constant we first need some preparations.

Definition 5. A *graph height function* on a graph $G = (V, E)$ with respect to a given origin-vertex $o \in V$ is a pair $(h, \mathcal{H})$ such that

(a) $h : V \to \mathbb{Z}$, and $h(o) = 0$,

(b) $\mathcal{H} \leq AUT(G)$ acts quasi-transitively on G and h is $\mathcal{H}$-difference-invariant in the sense that

$$h(\alpha v) - h(\alpha u) = h(v) - h(u) \quad \text{for all } \alpha \in \mathcal{H}, \ u, v \in V,$$

(c) for $v \in V$, there exist $u, w \in V$ neighbours of v such that $h(u) < h(v) < h(w)$.

A graph height function $(h, \mathcal{H})$ is called *unimodular* if the action of $\mathcal{H}$ on G is unimodular, i.e. if

$$|Stab_u^{\mathcal{H}} v| = |Stab_v^{\mathcal{H}} u| \quad \text{for any } v \in V, \ u \in \mathcal{H}v,$$

where $Stab_u^{\mathcal{H}} v = \{\gamma v \mid \gamma \in Stab_u^{\mathcal{H}}\}$.

Definition 6. Assume that $G = (V, E)$ is an infinite quasi-transitive graph with graph height function $(h, \mathcal{H})$. Let $v \in V$ and $\pi = (v_0, v_1, \ldots, v_n) \in \Sigma_n(v)$. We call π a *bridge* if

$$h(v_0) < h(v_i) \leq h(v_n) \quad \text{for all } \ 1 \leq i \leq n.$$

For a given integer $n \geq 0$ the set of n-step bridges starting in v is denoted by $B_n(v)$ and its cardinality by $b_n(v)$.

The following theorem by Grimmett and Li [13] serves as the definition of the bridge constant of a quasi-transitive graph with a given height function.

© Springer Fachmedien Wiesbaden GmbH, part of Springer Nature 2018
C. Lindorfer, *The Language of Self-Avoiding Walks*, BestMasters,
https://doi.org/10.1007/978-3-658-24764-5_3

Theorem 5. *Let $G = (V, E)$ be an infinite, quasi-transitive graph possessing a graph height function $(h, \mathcal{H})$. Then there exists $\beta = \beta(G, h, \mathcal{H}) \in \mathbb{R}$, called the bridge constant such that*

$$\beta = \lim_{n \to \infty} b_n(v)^{\frac{1}{n}} \quad \text{for all} \quad v \in V.$$

Proof. We will only prove the case where G is transitive.

As G is transitive $b_n := b_n(v)$ is independent of the choice of v for all integers $n \geq 0$. Note that $b_n \geq 1$ for all $n \geq 0$ as each vertex has a neighbour of larger height.

The concatenation of a bridge $\pi^{(1)} = (v_0^{(1)}, v_1^{(1)}, \ldots, v_m^{(1)})$ of length $m \geq 0$ starting at $v \in V$ and ending at some $w \in V$ and a second bridge $\pi^{(2)} = (v_0^{(2)}, v_1^{(2)}, \ldots, v_n^{(2)})$ of length $n \geq 0$ starting at $w \in V$ is again a SAW as $h(v_i^{(1)}) \leq h(w)$ for all $0 \leq i \leq m$ and $h(w) < h(v_j^{(2)})$ for all $1 \leq j \leq n$ and therefore also a bridge as $h(v_0^{(1)}) < h(u) \leq h(v_m^{(2)})$ for all inner vertices u of the concatenated walk. Therefore we have

$$b_m b_n \leq b_{m+n} \quad \text{for all integers } m, n \geq 0.$$

This is equivalent to $-\log(b_n)$ being a subadditive sequence and by using Lemma 1 we get that

$$\lim_{n \to \infty} \frac{-\log b_n}{n} = \inf_{n \geq 1} \frac{-\log b_n}{n} \in [-\infty, 0].$$

We can therefore define the bridge constant β by

$$\beta := \lim_{n \to \infty} b_n^{1/n} \leq \mu < \infty.$$

The upper bound μ follows from the simple observation that $b_n(v) \leq \sigma_n(v)$ holds for every $v \in V$. $\qquad \square$

It is now possible to formulate the Bridge Theorem by Grimmett and Li [13], which shows that the bridge constant with respect to unimodular height functions is equal to the connective constant and also that the bridge constant does not depend on the choice of the (unimodular) height function.

Theorem 6. *Let $G = (V, E)$ be an infinite, quasi-transitive graph possessing a unimodular graph height function $(h, \mathcal{H})$. Then $\beta(G, h, \mathcal{H}) = \mu(G)$.*

We give an example showing that in the Bridge theorem (Theorem 6) the assumption of the height function being unimodular is indeed necessary.

This example was given without any details by Grimmett and Li in [13]. We will need the following definition about ends of trees.

Definition 7. Let $T = (V, E)$ be an infinite tree. A *ray* starting at $v \in V$ is an infinite sequence $\pi = (v = v_0, v_1, v_2, \dots)$ with $v_i \in V$ and $\{v_i, v_{i+1}\} \in E$ for all $i \geq 0$, where no vertex appears more than once in π.
We say that two rays of T are equivalent if they share all but finitely many vertices. This defines an equivalence relation on the set of all rays of T and we call the equivalence classes of this relation the *ends* of the tree.
We say that $\gamma \in AUT(T)$ fixes an end ξ of T, if $\gamma\xi = \xi$, i.e., for a ray $\pi \in \xi$, also $\gamma\pi \in \xi$.

Example 3. Let $\mathbb{T}_3 = (V, E)$ be the regular tree with vertex degree 3. From Example 1 we already know that $\mu(\mathbb{T}_3) = 2$. We will define a non-unimodular height function on $\mathbb{T}_3$.

Let $v_0 \in V$ be a given vertex, $\pi_0 = (v_0, v_1, v_2, \dots)$ be a ray in $\mathbb{T}_3$ starting at v_0 and ξ be the end of $\mathbb{T}_3$ represented by π_0.
We denote by π_v the ray starting at $v \in V$ and representing ξ and call its vertices ancestors of v (v is an ancestor of itself).
Let $h : V \to \mathbb{Z}$ defined by $h(v_k) = k$ for all $k \geq 0$ and $h(v) = h(v_j) - d(v, v_j)$ where j is such that v_j is an ancestor of v as shown in Figure 3.1. Obviously the definition does not depend on the choice of the ancestor.

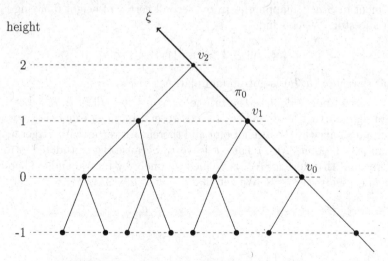

Figure 3.1: Non-unimodular height function on $\mathbb{T}_3$.

For $\mathcal{H}$ we pick the subgroup of $AUT(\mathbb{T}_3)$ fixing ξ. Then $\mathcal{H}$ acts transitively on $\mathbb{T}_3$: Let $u, v \in V$ and $w \in V$ be a common ancestor of u and v (exists as π_u and π_v are equivalent and therefore share all but finitely many vertices). We pick for σ_u an element of $AUT(\mathbb{T}_3)$ sending π_u to π_w (there exist infinitely many of them). Then clearly σ_u is in $\mathcal{H}$ as π_u and π_w both represent ξ. In the same way we get $\sigma_v \in \mathcal{H}$ and therefore the automorphism $\sigma_u \sigma_v^{-1} \in \mathcal{H}$ sending u to v.

The pair $(h, \mathcal{H})$ defines a graph height function on $\mathbb{T}_3$: Clearly, (a) and (c) of Definition 5 are fulfilled. For (b) let $\gamma \in \mathcal{H}$ and $u, v \in V$ be given. Let $i, j \geq 0$ such that $v_i \in \pi_0$ is a common ancestor of u and v and $v_j \in \pi_0$ is a common ancestor of γu and γv. Then $\pi_{v_i} \cap \pi_{\gamma^{-1} v_j}$ is non-empty, as both rays represent ξ. Also $\gamma^{-1} \pi_{v_j}$ is a ray starting in $\gamma^{-1} v_j$ and representing ξ, therefore equals $\pi_{\gamma^{-1} v_j}$. Let $l \geq 0$ such that $v_l \in \pi_{v_i} \cap \gamma^{-1} \pi_{v_j}$. Then v_l is a common ancestor of u and v and $\gamma(v_l) \in \pi_0$ is a common ancestor of γu and γv. Using that $h(u) = h(v_l) - d(u, v_l)$ and $h(\gamma u) = h(\gamma v_l) - d(\gamma u, \gamma v_l)$ and the similar statements for v we get

$$h(\gamma u) - h(\gamma v) = -d(\gamma u, \gamma v_l) + d(\gamma v, \gamma v_l) = -d(u, v_l) + d(v, v_l) = h(u) - h(v).$$

The height function $(h, \mathcal{H})$ is not unimodular. We consider $Stab_{v_i}^{\mathcal{H}}$ and note that every $\gamma \in \mathcal{H}$ fixing v_i also needs to fix π_{v_i} as γ maps rays onto rays and would otherwise map ξ to a different end of $\mathbb{T}_3$. Also, there is an element in $Stab_{v_1}^{\mathcal{H}}$ mapping v_0 to the second vertex of height 0 having v_1 as an ancestor. We conclude

$$|Stab_{v_1}^{\mathcal{H}} v_0| \geq 2 > 1 = |\{v_1\}| = |Stab_{v_0}^{\mathcal{H}} v_1|$$

and therefore $(h, \mathcal{H})$ is not unimodular.

We will now calculate the bridge constant $\beta = \beta(\mathbb{T}_3, h, \mathcal{H})$. Let ν be a bridge starting at v_0. Then ν needs to end in π_0 as otherwise it would contain an ancestor of its endpoint and therefore a vertex with bigger height than its endpoint. Also ν cannot leave π_0 because the considered graph is a tree and therefore a SAW is defined uniquely by its two ends. Therefore $b_n(v_0) = 1$ for all $n \geq 0$. We get that $1 = \beta \neq \mu = 2$.

3.2. Bridges on strips of the integer lattice

The goal in this section is to calculate the connective constant of the ladder graph $\mathbb{L}$ by using the Bridge Theorem.

Example 4. Let $\mathbb{L} = (V, E)$ where $V = \mathbb{Z} \times \{0, 1\}$ and two vertices u, v are connected by an edge if and only if $|u - v| = 1$, where the absolute value is the usual euclidean norm in $\mathbb{R}^2$.

We pick $(0, 0)$ as origin, $h : V \to \mathbb{Z}$, $(x, y) \mapsto x$ and for $\mathcal{H}$ the set of all horizontal translations $(x, y) \mapsto (x + k, y)$ for some $k \in \mathbb{Z}$. Then the pair $(h, \mathcal{H})$ is a graph height function and it is unimodular as $Stab_v^{\mathcal{H}} = \{id_V\}$ (the identity map from V to V) for all $v \in V$.

For counting b_n, the number of bridges of length n starting in $(0, 0)$, we use ordinary generating functions. Let the generating function of the number of bridges be

$$F_B(t) = \sum_{n=0}^{\infty} b_n t^n.$$

Using the same method as in Remark 5 and the Bridge Theorem we get for the radius of convergence R of $F_B(t)$ that $\mu = \beta = 1/R$.

We will use the same kind of "linguistic method" as Zeilberger in [22], who counted the number of n-step SAWs on $\mathbb{L}$.

Start at $(0, 0)$ and denote a step right by r, a step up or down (depending on the current position) by s and a step left by l (we will not need these). Then every bridge π has to start with a finite positive number of steps r, as all vertices of π except its starting vertex need to have positive height. Then we can only have a step s, and afterwards r again (we cannot have l, otherwise π could not end at a point with maximal height). This implies that every bridge is of the form L^*I, where L^* denotes a concatenation of any non-negative number of walks of type L and

(a) L is a walk of the form $r^i s$, $i \geq 1$,

(b) I is a walk of the form r^i, $i \geq 0$.

This implies that the generating function of L is

$$F_L(t) = t^2 + t^3 + t^4 + \cdots = \frac{t^2}{1 - t} \tag{3.1}$$

and the generating function of I is

$$F_I(t) = 1 + t + t^2 + \cdots = \frac{1}{1 - t}. \tag{3.2}$$

Using (3.1) we get for the generating function $F_{L^*}(t)$ of L^*, which clearly is the sequence class of L as defined in Remark 4:

$$F_{L^*}(t) = 1 + F_L(t) + F_L(t)^2 + \cdots = \frac{1}{1 - F_L(t)}. \tag{3.3}$$

(3.2) and (3.3) then give for the generating function of bridges

$$F_B(t) = F_{L^*}(t) F_I(t) = \left(\frac{1-t}{1-t-t^2} \right) \left(\frac{1}{1-t} \right) = \frac{1}{1-t-t^2}.$$

This function has exactly two poles at $t_{1,2} = (-1 \pm \sqrt{5})/2$, where the one with plus has the smaller absolute value. We conclude that

$$\mu(\mathbb{L}) = \frac{1}{R} = \frac{2}{\sqrt{5}-1} = \frac{1+\sqrt{5}}{2}.$$

As we have already seen, the concatenation of two bridges is again a bridge. So there are bridges which can be decomposed into shorter bridges. This observation leads to the following definition.

Definition 8. We call an n-step bridge π *irreducible* if it cannot be decomposed into smaller bridges, i.e. there are no bridges π_1, π_2 of length $< n$ such that π is the concatenation of π_1 and π_2.

Lemma 2. *Let G be a quasi-transitive graph and $(h, \mathcal{H})$ be a graph height function on G. A bridge $\pi = (v_0, v_1, \ldots v_n)$ is irreducible if and only if there is no integer k with $0 < k < n$ such that*

$$h(v_i) \le h(v_k) < h(v_j) \quad \text{for all } 0 \le i \le k, \ k+1 \le j \le n. \tag{3.4}$$

Proof. Let $\pi = (v_0, v_1, \ldots v_n)$ be a bridge and assume there is an integer k with $0 < k < n$ such that (3.4) holds. Using the assumption and that π is a bridge, we get

$$h(v_0) < h(v_i) \le h(v_k) < h(v_j) \le h(v_n) \quad \text{for all } 1 \le i \le k, k+1 \le j \le n.$$

We conclude that π is not irreducible as it can be decomposed into the two bridges $\pi_1 = (v_0, \ldots v_k)$ and $\pi_2 = (v_k, \ldots v_n)$.

Conversely suppose that there is an integer k with $0 < k < n$ such that π can be decomposed into the bridges $\pi_1 = (v_0, \ldots v_k)$ and $\pi_2 = (v_k, \ldots v_n)$. Then by the definition of bridges, (3.4) holds. $\square$

Corollary 1. *Every bridge can be uniquely decomposed into a finite number of irreducible bridges.*

Proof. Let $\pi = (v_0, v_1, \ldots v_n)$ be a bridge and denote by K the set of all k such that $0 < k < n$ and (3.4) holds. Decomposing π at the $|K|$ vertices v_k for $k \in K$ gives a decomposition into $|K| + 1$ bridges.

Let $L \subset \{1, \ldots, n-1\}$ be a set of vertices such that we can decompose π at the $|L|$ vertices v_l for $l \in L$ into $|L| + 1$ irreducible bridges. Then for every $l \in L$ (3.4) holds. We get $L \subset K$ and by the irreducibility of the resulting bridges also $L = K$. Therefore our decomposition at the vertices v_k for $k \in K$ is the unique decomposition into irreducible bridges. $\square$

Dangovski calculated the generating functions of SAWs on the integer strip $SQ_{-1,1} = \mathbb{Z} \times \{-1, 0, 1\}$ in [6]. To achieve this he used the same "linguistic" method as Zeilberger in [22]. He divided the walks into subwalks of certain types and first solved some smaller problems. Clearly there are a lot of cases to be distinguished and a lot of work has to be done. Although there were some mistakes in his proof, he got the following correct result:

Theorem 7. *Let $W(t)$ be the generating function of all SAWs on $SQ_{-1,1}$ starting in $(0,0)$. Then*

$$W(t) = \frac{N(t)}{D(t)}$$

where

$$N(t) = 4t^{22} + 4t^{21} + 4t^{20} - 4t^{18} + 26t^{17} + 24t^{16} + 3t^{15} - $$
$$38t^{14} - 16t^{13} + 32t^{12} + 31t^{11} - 10t^{10} - 35t^9 - 11t^8 + $$
$$21t^7 + 14t^6 - 2t^5 - 10t^4 - 3t^3 + 2t^2 + 3t + 1$$

and

$$D(t) = (2t^6 + 2t^5 + t^4 + 2t^3 + t - 1)(t^4 + 1)^2(2t^2 - 1)^2(t^2 + t + 1)(t - 1).$$

Moreover

$$\mu(SQ_{-1,1}) = 1/v_{min} \approx 1.914626790719$$

where v_{min} is the root of $D(t)$ having minimal absolute value.

A much shorter approach to get the connective constant of $SQ_{-1,1}$ is again by calculating the generating function of bridges. Our method here uses some ideas from the work of Dangovski and Lalov in [5], but the combinatorial part is different.

Example 5. Consider $SQ_{-1,1} = \mathbb{Z} \times \{-1, 0, 1\}$ together with the unimodular graph height function $(h, \mathcal{H})$ where $h : V \to \mathbb{Z}$, $(x, y) \mapsto x$ and $\mathcal{H}$ the set of all horizontal translations $(x, y) \mapsto (x + k, y)$ for any $k \in \mathbb{Z}$. Similar to Example 4, we denote a step right by r, left by l, up by u and down by d. We will split any bridge π on $SQ_{-1,1}$ starting at $(0, 0)$ in its irreducible bridges. By Corollary 1 this procedure is unique.

We call the line $\mathbb{Z} \times \{0\}$ inner line and the lines $\mathbb{Z} \times \{\pm 1\}$ outer lines and divide the set of irreducible bridges into the following 4 types:

(a) I_I: starts at inner line and ends at inner line.

(b) I_O: starts at inner line and ends at outer line.

(c) O_O: starts at outer line and ends at outer line.

(d) O_I: starts at outer line and ends at inner line.

The following Table 3.1 lists the form of the elements and its generating function for each type. For types starting with O we assume the bridge starts at the bottom line.

	Form of bridge	Generating function
I_I	r	$F_{I_I}(t) = t$
I_O	ru, rd	$F_{I_O}(t) = 2t^2$
O_O	r, $r^{i+1}ul^iur^i$ for $i \geq 0$	$F_{O_O}(t) = t + \frac{t^3}{1-t^3}$
O_I	ru	$F_{O_I}(t) = t^2$

Table 3.1: Irreducible bridges on $SQ_{-1,1}$ and their generating functions.

We want to count bridges starting in $(0, 0)$ at the inner line. The idea is to build the bridge by concatenating all possible irreducible bridges. First we build all minimal bridges starting and ending at the inner line and get the class A (minimal here means that we cannot decompose it into two bridges of this type). Its elements are of the form

$$A : \begin{cases} I_I \\ I_O(O_O)^*O_I. \end{cases}$$

Therefore its generating function $F_A(t)$ satisfies

$$F_A(t) = F_{I_I}(t) + F_{I_O}(t)\frac{1}{1 - F_{O_O}(t)}F_{O_I}(t) = \frac{t - t^2 + t^5 - 2t^7}{1 - t - 2t^3 + t^4}.$$

Now we get for the class B of all bridges in $SQ_{-1,1}$ the characterization

$$B : \begin{cases} A^* \\ A^* I_O (O_O)^*. \end{cases}$$

Here the first line describes all bridges ending at the middle line and the second line describes all bridges ending at an outer line. The generating function $F_B(t)$ corresponding to B satisfies

$$
\begin{aligned}
F_B(t) &= \frac{1}{1 - F_A(t)} \left(1 + F_{I_O}(t) \frac{1}{1 - F_{O_O}(t)} \right) \\
&= \frac{1 - t + 2t^2 - 2t^3 + t^4 - 2t^5}{(1 - t - 2t^3 - t^4 - 2t^5 - 2t^6)(1 - t)}.
\end{aligned}
$$

By Pringsheim's Theorem the smallest positive root of the denominator must have the smallest absolute value of all roots of the denominator. By taking the reciprocal of this root and then using Theorem 6 we get

$$\mu(SQ_{-1,1}) \approx 1.91462679.$$

Obviously $SQ_{-1,1}$ is not a transitive graph, as vertices on the inner line have degree 4 while the vertices at the outer line have degree 3. It seems natural to also study the transitive graph $SQ^c_{-1,1}$ resulting from $SQ_{-1,1}$ when adding all edges $\{(x, -1), (x, 1)\}$ for $x \in \mathbb{Z}$. In the next example we calculate the generating function of bridges and the connective constant of $SQ^c_{-1,1}$.

Example 6. We will use the unimodular height function defined in Example 5 and again denote steps up, down left and right by u, d, l and r respectively. The difference is that we can now also go up from a vertex $(x, 1)$ to get to the vertex $(x, -1)$ and down from $(x, -1)$ to reach $(x, 1)$. Furthermore we will again build all bridges starting at $(0,0)$ by concatenating irreducible bridges. Using Lemma 2 it is not difficult to show that every irreducible bridge is of one of the forms listed in table 3.2:

We start by building the class A of all bridges containing exactly one irreducible bridge of type S which has to be at the end. Previous to the bridge of type S, there can be arbitrary many walks of type L, which can have arbitrary many I-bridges between them. We get the characterization

$$A : (I^* L)^* I^* S.$$

	Form of bridge	Generating function
I	r	$F_I(t) = t$
L	ru, rd	$F_L(t) = 2t^2$
S	$r^{i+1}ul^iur^i, r^{i+1}dl^idr^i$ for $i \geq 0$	$F_S(t) = \frac{2t^3}{1-t^3}$

Table 3.2: Irreducible bridges on $SQ^c_{-1,1}$ and their generating functions.

With this we can build any bridge on $SQ^c_{-1,1}$ starting in $(0,0)$ by concatenating arbitrary many bridges of type A and then maybe adding some final part not containing any walk of type S. Thus we get for the class of all bridges B the characterization

$$B : A^*(I^*L)^*I^*.$$

The resulting generating functions corresponding to the classes A and B are

$$F_A(t) = \frac{1}{(1 - \frac{1}{1-F_I(t)}F_L(t))(1 - F_I(t))}F_S(t)$$

$$= \frac{2t^3}{(1+t+t^2)(1-2t)(1+t)(1-t)},$$

$$F_B(t) = \frac{1}{(1 - F_A(t))(1 - \frac{1}{1-F_I(t)}F_L(t))(1 - F_I(t))}$$

$$= \frac{1-t^3}{1-t-2t^2-3t^3+t^4+2t^5}.$$

Using Theorem 6 we get the connective constant as the reciprocal of the smallest modulus root of the denominator of $F_B(t)$:

$$\mu(SQ^c_{-1,1}) \approx 2.28965789.$$

3.3. Convergence of connective constants

Consider again the lattice $\mathbb{Z}^2$ and the sub-lattices $SQ_{0,L} = \mathbb{Z} \times \{0, \ldots, L\}$ of $\mathbb{Z}^2$. Following the approach of Beffara and Huynh in [2] we will now prove that the sequence $\mu(SQ_{0,L})$ converges to $\mu(\mathbb{Z}^2)$ for L going to infinity. As in Example 5 we take for any given L as unimodular height function on $SQ_{0,L}$ the pair $(h, \mathcal{H})$ where $h : V \to \mathbb{Z}$, $(x,y) \mapsto x$ and $\mathcal{H}$ the set of all horizontal translations $(x,y) \mapsto (x+k, y)$ for $k \in \mathbb{Z}$.

Lemma 3. *For integers $L, n \geq 1$ let $B_n^{(L)}$ be the set of bridges in $SQ_{-L,L}$ starting at the origin $(0,0)$ and ending in some (x, y) with $y \geq 0$ and let $b_n^{(L)}$ be its cardinality $|B_n^{(L)}|$. Then*

$$\forall L, n, k \geq 1 : b_{kn}^{(3L)} \geq (b_n^{(L)})^k. \tag{3.5}$$

Proof. We will use induction on k to prove the following Claim:

For a sequence of k given bridges $\pi_1, \ldots, \pi_k \in B_n^{(L)}$ we can concatenate the bridges or their reflection on the line $\mathbb{Z} \times \{0\}$ such that we get a new bridge $\pi_{1,k} \in B_{kn}^{(3L)}$ ending in some (x_k, y_k) with $0 \leq y_k \leq 2L$.

Clearly the claim holds for $k = 1$: $\pi_{1,1} = \pi_1$ is in $B_n^{(L)}$, therefore also in $B_n^{(3L)}$ and is ending in some (x_1, y_1) with $0 \leq y_1 \leq 2L$. Assume the claim holds for $k-1$. Then we can concatenate the bridges $\pi_1, \ldots, \pi_{k-1} \in B_n^{(L)}$ to get a bridge $\pi_{1,k-1} \in B_{(k-1)n}^{(3L)}$ ending in some (x_{k-1}, y_{k-1}) with $0 \leq y_{k-1} \leq 2L$. Let $\pi_k \in B_n^{(L)}$ end at (x, y). We distinguish 2 cases:

(a) If $y_{k-1} \leq L$, concatenate $\pi_{1,k-1}$ and π_k to get $\pi_{1,k}$, which is a bridge in $B_{kn}^{(3L)}$, its end-vertex (x_k, y_k) satisfies

$$0 \leq y_{k-1} \leq y_k = y_{k-1} + y \leq y_{k-1} + L \leq 2L.$$

(b) If $y_{k-1} > L$, concatenate π_1^{k-1} and the reflection of π_k on $\mathbb{Z} \times \{0\}$ to get π_1^k, which is a bridge in $B_{kn}^{(3L)}$, its end-vertex (x_k, y_k) satisfies

$$0 < y_{k-1} - L \leq y_{k-1} - y = y_k \leq y_{k-1} \leq 2L.$$

This concludes our induction. An example for the construction is shown in Figure 3.2, where the arrow heads represent the end of the four concatenated bridges.

In the above construction we cannot get a bridge in $B_{kn}^{(3L)}$ more than once. This fact together with the above claim proves the statement (3.5) of the Lemma. $\square$

Theorem 8. *Let $L \geq 1$ be an integer and $\mu(SQ_{0,L})$ and $\mu(\mathbb{Z}^2)$ be the connective constants of the integer strips $SQ_{0,L}$ and the integer lattice $\mathbb{Z}^2$. Then*

$$\lim_{L \to \infty} \mu(SQ_{0,L}) = \mu(\mathbb{Z}^2). \tag{3.6}$$

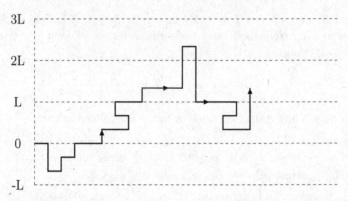

Figure 3.2: Concatenation of four bridges in $B_n^{(L)}$, result is in $B_{4n}^{(3L)}$.

Proof. Let B_n be the set of all n-step bridges on $\mathbb{Z}^2$ starting in $(0,0)$, $b_n = |B_n|$ its cardinality and b_n' be the number of bridges in B_n ending in some (x, y) with $y \geq 0$. Also let $b_n^{(L)}$ defined as in Lemma 3. By Theorem 6 we know

$$\lim_{n \to \infty} (b_n)^{1/n} = \mu(\mathbb{Z}^2). \tag{3.7}$$

Using that every bridge on $\mathbb{Z}^2$ either has some endpoint (x, y) for $y \geq 0$ or its reflection on $\mathbb{Z} \times \{0\}$ does, we get

$$b_n \leq 2b_n' \leq 2b_n,$$

which together with 3.7 proves

$$\forall L \geq 1 : \lim_{n \to \infty} (b_n')^{1/n} = \mu(\mathbb{Z}^2). \tag{3.8}$$

The same argument as above used on $SQ_{-L,L}$ gives us

$$\lim_{n \to \infty} (b_n^{(L)})^{1/n} = \mu(SQ_{-L,L}). \tag{3.9}$$

Every bridge counted in b_L' contains only vertices (x, y) with $-L \leq y \leq L$ as it has length L. Therefore it is also counted in $b_L^{(L)}$. This gives

$$\forall L \geq 1 : b_L^{(L)} = b_L'. \tag{3.10}$$

Let $\epsilon > 0$ be given and using (3.8) take n_0 such that

$$|(b_n')^{1/n} - \mu(\mathbb{Z}^2)| \leq \epsilon \quad \text{for all} \quad n \geq n_0. \tag{3.11}$$

By Lemma 3 and (3.10) we know that

$$(b_{kn_0}^{(3n_0)})^{\frac{1}{kn_0}} \geq (b_{n_0}^{(n_0)})^{\frac{1}{n_0}} = (b'_{n_0})^{\frac{1}{n_0}}. \tag{3.12}$$

Using (3.9) we get that

$$\lim_{k \to \infty} (b_{kn_0}^{(3n_0)})^{\frac{1}{kn_0}} = \mu(SQ_{-3n_0, 3n_0}). \tag{3.13}$$

Now sending k to infinity in (3.12) and using (3.11), (3.13) and that $\mu(SQ_{0,L})$ is strictly increasing in L we obtain

$$\mu(SQ_{0,L}) > \mu(SQ_{-3n_0, 3n_0}) \geq \mu(\mathbb{Z}^2) - \epsilon \quad \text{for all } L > 6n_0.$$

This together with $\mu(SQ_{0,L}) \leq \mu(\mathbb{Z}^2)$ for all $L \geq 0$ proves the claim (3.6).

$\square$

4. Self-avoiding walks on one-dimensional lattices

4.1. Configurations and Shapes

In this chapter we will mostly follow the approach of Alm and Janson in [1], who showed that the generating function of SAWs on one-dimensional lattices can be explicitly expressed in terms of a rational function and that therefore the connective constant is algebraic as the reciprocal of a root of the denominator. We start by defining lattices and their dimension in a convenient way.

Definition 9. For an integer $d \geq 1$, a *d-dimensional lattice* is an infinite graph G such that there is a group $\Gamma \leq AUT(G)$ of translations with $\Gamma \cong \mathbb{Z}^d$ and the number of orbits of Γ acting on G is finite. In other words, the vertices may be represented by $\mathbb{Z}^d \times F$, where $F = \{1, 2, \ldots, |F|\}$ is a finite set and two vertices (m, a) and (n, b) are connected by an edge if and only if $n - m \in E_{a,b}$, where for each pair $(a, b) \in F^2$, $E_{a,b}$ is a finite subset of $\mathbb{Z}^d$. We say that a vertex (m, a) has *longitude m* and *latitude a*.

For a better understanding of the definition we will give a simple example of a lattice of dimension 1.

Example 7. Consider the triangular strip TRI_2 as in Figure 4.1. We will state two different ways for defining the sets F and E_{ab} from Definition 9.

(a) We set $F_1 = \{1, 2\}$ and $V = \mathbb{Z} \times F_1$. Then a possible set of edges of TRI_2 is given by $E_{1,1} = E_{2,2} = \{-1, 1\}$, $E_{1,2} = \{0, -1\}$, $E_{2,1} = \{0, 1\}$. As we only consider undirected graphs, it is clear that $E_{a,b} = -E_{b,a}$, where the minus is applied on each element of the set. Therefore the lattice is defined by the sets $E_{a,b}$ where $a \leq b$.

(b) Let $F_2 = \{1, 2, 3, 4\}$ and $V = \mathbb{Z} \times F_2$. Then we can define the set of edges by $E_{1,1} = E_{2,2} = E_{3,3} = E_{4,4} = \emptyset$, $E_{1,2} = E_{2,3} = E_{3,4} = \{0\}$, $E_{1,3} = E_{2,4} = \{0, -1\}$ and $E_{1,4} = \{-1\}$.

Clearly there are infinitely many other ways of describing TRI_2 in the same way as above. It is also possible to get a set F with $|F| = 1$, but then $E_{a,b} \subset \{-1, 0, 1\}$ cannot be satisfied anymore.

© Springer Fachmedien Wiesbaden GmbH, part of Springer Nature 2018
C. Lindorfer, *The Language of Self-Avoiding Walks*, BestMasters,
https://doi.org/10.1007/978-3-658-24764-5_4

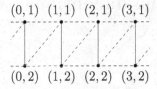 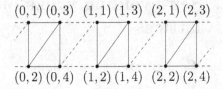

Figure 4.1: Two ways of defining F and $V_{a,b}$ for TRI_2.

From now on we will only consider one-dimensional lattices. As we have seen in the example, there are many different ways to represent a lattice. The following lemma will grant us the existence of a particularly beneficial type of representation.

Lemma 4. *For every one-dimensional lattice G there is a representation such that*

(a) *$E_{a,b} \subset \{-1, 0, 1\}$ for all $a, b \in F$, i.e., edges occur only between successive longitudes and*

(b) *each vertex v of longitude $i \in \mathbb{Z}$ is connected to a vertex w of longitude $i + 1$ by a walk in which all vertices besides w have longitude i.*

Proof. Consider a representation of a one-dimensional lattice as in Definition 9 with vertices (n, a) where $n \in \mathbb{Z}$, $a \in F$ and let

$$H = \max_{a,b \in F} \max_{k \in E_{a,b}} |k|.$$

We want to group H successive longitudes into a single new longitude. To achieve this we can use the following map φ from $\mathbb{Z} \times F$ to $\mathbb{Z} \times \{1, 2, \ldots, H|F|\}$:

$$\varphi((Hn + k, a)) = (n, k|F| + a) \quad \text{for } 0 \leq k \leq H - 1,\ a \in F$$

giving us a new representation with $H|F|$ latitudes. Let $(Hn + k, a)$ and $(Hm + l, b)$ be two vertices connected by an edge in the original lattice. Then $|Hn + k - (Hm + l)| \leq H$ which implies that $|n - m| \leq 1$. So the new lattice $\varphi(G)$ satisfies $E_{a,b} \subset \{-1, 0, 1\}$.

Now suppose that we have a representation of G with $E_{a,b} \subset \{-1, 0, 1\}$. Consider the subgraph G_0 of G induced by the vertices of longitude 0. We call the connected components of G_0 clusters. Suppose there is a cluster C not directly connected to a vertex of longitude 1 and let I be the set of latitudes of all vertices in C. Then C is directly connected to a vertex $(-1, a)$ with $a \in F \setminus I$ because G is connected and $E_{a,b} \subset \{-1, 0, 1\}$. We use the map ψ from $\mathbb{Z} \times F$ to $\mathbb{Z} \times F$ defined by

$$\psi((n, a)) = \begin{cases} (n-1, a) & \text{if } a \in I, \\ (n, a) & \text{otherwise} \end{cases}$$

to reduce the longitude of C (and its translates) by one. The new graph has one cluster less in G_0. Thus, by choosing a representation with a minimal number of clusters every cluster of longitude 0 has to be connected to a vertex with longitude 1.   □

Let G be a one-dimensional lattice and choose a representation as provided by Lemma 4. We will describe G in the following way (see Figure 4.2). We divide the graph into segments, each of them consisting of all vertices of a single longitude, all edges connecting two vertices in the segment and all edges connecting the segment and the preceding segment.

Furthermore we will divide each segment into two parts, a section, comprising only the connections (edges) to the preceding segment, and a hinge, containing all vertices of the segment and all edges between these vertices.

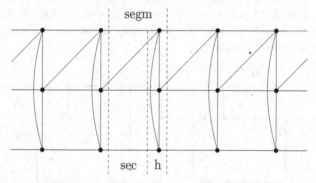

Figure 4.2: Segment (segm), section (sec) and hinge (h) of a lattice, following [1].

Let π be a SAW connecting two vertices S (start) and E (end) on the one-dimensional lattice G. Then we can see π as a sequence of directed edges, ordered by its appearance in π. The appearance of π in a particular section of G is called a configuration. So a configuration is a sequence of directed edges ordered by its appearance in π. Let $\mathcal{C}$ be the set of all possible configurations that may appear in a SAW on G. For convenience, we include two different empty configurations ϕ_L and ϕ_R in $\mathcal{C}$.

Similarly the appearance of the SAW π in a particular hinge is called a shape, which is a sequence of vertices and their incoming and outgoing edges of π. Here we order the vertices by their appearance in π.

We note that the shape of a hinge is not necessarily determined by the configurations of the adjacent sections. On the other hand a shape completely determines the configurations on both sides.

We call a configuration even (odd) if it contains an even (odd) number of edges. Furthermore it is called left or right according to the direction of the first edge. Note that if S lies to the left of E, the configurations to the left of S are even left, the configurations between S and E are odd and those to the right of E are even right. We think of ϕ_L (ϕ_R) as an empty configuration to the left (right) of π and therefore define it to be even left (right). So we have a partition of all configurations into 4 sets

$$\mathcal{C} = \mathcal{C}_{EL} \cup \mathcal{C}_{OL} \cup \mathcal{C}_{OR} \cup \mathcal{C}_{ER}, \tag{4.1}$$

where EL, OL, OR and ER stand for even left, odd left, odd right and even right. A Self-avoiding walk on the one-dimensional lattice $SQ_{0,3}$ and the types for each configuration are shown in Figure 4.3. Note that no configurations of type OL occur in this example. We will show in the proof of Lemma 5 that this is the case, whenever the start S lies to the left of the end E. On the ladder graph $\mathbb{L}$ there are already 10 possible configurations and 38 possible shapes, which are shown in Figure 4.5 and Figure 4.6 sorted by their type.

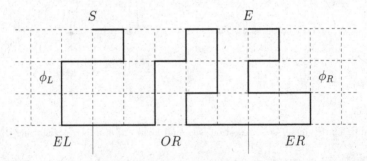

Figure 4.3: A SAW connecting S and E, including types of occurring configurations.

We say that the shape s correctly connects the configurations c_1 and c_2, if the edges leaving s to the left are exactly the edges in c_1, their order is preserved and the same holds for the edges leaving s to the right and the edges of c_2.

We can describe a SAW π by a sequence of configurations c_i and shapes s_i of the form $(\phi_L = c_0, s_1, c_1, \ldots, s_m, c_m = \phi_R)$ where s_i correctly connects c_{i-1} and c_i for all $1 \le i \le n$. We will show next that the converse holds.

Lemma 5. *A correctly connected sequence $(\phi_L = c_0, s_1, c_1, \ldots, s_m, c_m = \phi_R)$ defines a self-avoiding walk π.*

Proof. We can distinguish 4 different types of shapes depending on whether they contain S, E, both or none of them. A shape containing neither S nor E can only connect two configurations of the same type (EL, OL, OR or ER) as the edges leave in the same order and quantity they enter. A shape containing S can only have a left configuration at the left and a right configuration at the right. A shape containing E but not S keeps the orientation (left or right) of its neighboured configurations but changes even to odd and vice versa. If a shape contains both E and S it has a configuration of type EL at the left and of type ER on the right. So the only possible sequences can be represented by the following diagram (Figure 4.4), where edges represent the transitions from one configuration to the next one and their label tells us if the corresponding shape contains S, E or both of them.

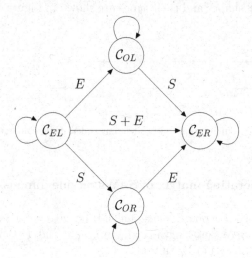

Figure 4.4: Possible connections of configurations, following [1].

Conversely, let $(\phi_L = c_0, s_1, c_1, \ldots, s_m, c_m = \phi_R)$ be a correctly connected sequence of configurations and shapes. We assumed that we start with $\phi_L \in \mathcal{C}_{EL}$ and end with $\phi_R \in \mathcal{C}_{ER}$. Therefore we need to start at the leftmost node in the diagram and get to the rightmost node. This shows us that the pattern defined by our correctly connected sequence contains exactly one starting point S and one endpoint E.

Suppose there exists a cycle C of edges in the pattern. Each configuration containing an edge of C needs to contain at least 2 edges of C as C is a cycle. So for each configuration we get an ordering on the edges of C contained in it. But then at some point a shape has to connect an edge of higher order with an edge of lower order, which contradicts the assumption of our sequence being correctly connected. Hence there are no cycles in the pattern.

We conclude that the pattern defined by our correctly connected sequence contains only a single walk which has to be self-avoiding and connects the two vertices S and E.   □

In the following example we will list all configurations and shapes that can occur in a SAW on the one-dimensional lattice $\mathbb{L}$.

Example 8. We consider the ladder $\mathbb{L}$ with vertices $\{(x,y) \mid x \in \mathbb{Z}, y \in \{0,1\}\}$ and pick the starting point $S = (0,0)$. The 10 possible configurations and the 38 shapes and their types are shown in Figure 4.5 and Figure 4.6 .

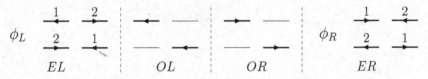

Figure 4.5: Possible configurations on $\mathbb{L}$, including order of edges and type, following [1].

4.2. The generating matrix of SAWs on one dimensional lattices

Let $G = (V, E)$ be a one dimensional lattice, $v \in V$ a fixed vertex, σ_n the number of SAWs of length n starting in v and $F(t)$ their ordinary generating function as in Definition 4.

Let $i, j \in \mathcal{C}$ be two configurations. We denote by $\eta(i)$ the number of edges in i and by $S(i,j)$ the set of all shapes correctly connecting i on the left with j on the right. For $s \in S(i,j)$ let $\nu(s)$ be the number of edges with both endpoints in s. We define the three $\mathcal{C} \times \mathcal{C}$ square matrices $H(t)$,

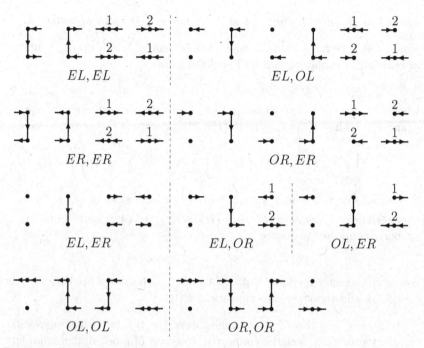

Figure 4.6: Possible shapes on $\mathbb{L}$, including order of vertices and type of configurations it can connect, following [1].

$V(t)$ and $M(t)$, whose entries are polynomials in the variable t. $M(t)$ will be called generating matrix of self-avoiding walks.

$$H = H(t) = (H_{ij})_{i,j \in \mathcal{C}} \quad \text{where} \quad H_{ij} = \begin{cases} t^{\eta(i)} & \text{for } j = i \\ 0 & \text{else} \end{cases},$$

$$V = V(t) = (V_{ij})_{i,j \in \mathcal{C}} \quad \text{where} \quad V_{ij} = \sum_{s \in S(i,j)} t^{\nu(s)} \quad \text{and}$$

$$M = M(t) = H(t)V(t).$$

Then the following theorem provides a connection of the generating function $F(t)$ and the matrices $H(t)$, $V(t)$ and $M(t)$.

Theorem 9. *Let $G = (V, E)$ be a one-dimensional lattice and $F(t)$, $H(t)$, $V(t)$ and $M(t)$ be defined as above. Then*

$$F(t) = V_{\phi_L \phi_R} + (VHV)_{\phi_L \phi_R} + (V(HV)^2)_{\phi_L \phi_R} + \cdots = (V(I - M)^{-1})_{\phi_L \phi_R},$$
$$(4.2)$$

where I is the identity matrix and A_{ij} denotes (i,j)-entry of matrix A.

Proof. Let π be a n-step SAW which has horizontal width m, i.e., contains m non-empty configurations and is of the form

$$(\phi_L = c_0, s_0, c_1, \ldots, s_m, c_{m+1} = \phi_R)$$

where the c_i are configurations and the s_i are shapes correctly connecting c_i and c_{i+1}. Then π appears in $(V(HV)^m)_{\phi_L \phi_R}$ as

$$t^{\nu(s_0)} \prod_{i=1}^{m} t^{\eta(c_i)} t^{\nu(s_i)} = \exp\left(\log(t) \left(\sum_{i=0}^{m} \nu(s_i) + \sum_{j=1}^{m} \eta(c_j) \right) \right) = t^n.$$

So every SAW π of length n contributes t^n to both sides of (4.2).

Conversely every term t^n in $(V(HV)^m)_{\phi_L \phi_R}$ of (4.2) corresponds to a properly connected sequence

$$(\phi_L = c_0, s_0, c_1, \ldots, s_m, c_{m+1} = \phi_R)$$

containing exactly n edges. But by Lemma 5 such a sequence corresponds to a SAW and therefore also appears in $F(t)$. $\square$

Applying this theorem we immediately get the following interesting corollary providing that the connective constant of a one-dimensional lattice is an algebraic number. Currently it is not known if this is also true for arbitrary graphs.

Corollary 2. *Let $G = (V, E)$ be a one-dimensional lattice. Then the generating function $F(t)$ corresponding to the number of SAWs on G is a rational function and the connective constant $\mu(G)$ is an algebraic number.*

Proof. $F(t)$ is an entry of the matrix $(V(I-M)^{-1})$, which has only rational functions as entries because V, I and M have only polynomials as entries. Therefore $F(t)$ can be seen as a rational function in its radius of convergence. By Theorem 4 we know that $F(t)$ has a singularity at $1/\mu(G)$. We conclude that $1/\mu(G)$ is a root of the denominator of $F(t)$ written as rational function and therefore algebraic. So also $\mu(G)$ is algebraic as reciprocal of an algebraic number. $\square$

Remark 6. We can use the partition of the index set $\mathcal{C}$ in (4.1) to get

$$H = \begin{pmatrix} H_{EL} & 0 & 0 & 0 \\ 0 & H_{OL} & 0 & 0 \\ 0 & 0 & H_{OR} & 0 \\ 0 & 0 & 0 & H_{ER} \end{pmatrix},$$

$$V = \begin{pmatrix} V_{EL,EL} & V_{EL,OL} & V_{EL,OR} & V_{EL,ER} \\ 0 & V_{OL,OL} & 0 & V_{OL,ER} \\ 0 & 0 & V_{OR,OR} & V_{OR,ER} \\ 0 & 0 & 0 & V_{ER,ER} \end{pmatrix} \quad \text{and}$$

$$M = \begin{pmatrix} M_{EL,EL} & M_{EL,OL} & M_{EL,OR} & M_{EL,ER} \\ 0 & M_{OL,OL} & 0 & M_{OL,ER} \\ 0 & 0 & M_{OR,OR} & M_{OR,ER} \\ 0 & 0 & 0 & M_{ER,ER} \end{pmatrix} =$$

$$= \begin{pmatrix} H_{EL}V_{EL,EL} & H_{EL}V_{EL,OL} & H_{EL}V_{EL,OR} & H_{EL}V_{EL,ER} \\ 0 & H_{OL}V_{OL,OL} & 0 & H_{OL}V_{OL,ER} \\ 0 & 0 & H_{OR}V_{OR,OR} & H_{OR}V_{OR,ER} \\ 0 & 0 & 0 & H_{ER}V_{ER,ER} \end{pmatrix}.$$

Example 9. We continue Example 8 and calculate the generating function of SAWs $F_{\mathbb{L}}(t)$ and the connective constant $\mu(\mathbb{L})$ of the ladder graph $\mathbb{L}$. The diagonal matrix H is determined by the 10 configurations in Figure 4.5. The sub-matrices of H introduced in Remark 6 are

$$H_{EL} = H_{ER} = \begin{pmatrix} 1 & 0 & 0 \\ 0 & t^2 & 0 \\ 0 & 0 & t^2 \end{pmatrix} \quad \text{and} \quad H_{OL} = H_{OR} = \begin{pmatrix} t & 0 \\ 0 & t \end{pmatrix}.$$

The matrix V is given by the 38 shapes in Figure 4.6. Its submatrices as defined in Remark 6 are

$$V_{EL,EL} = V_{ER,ER}^T = \begin{pmatrix} 0 & t & t \\ 0 & 1 & 0 \\ 0 & 0 & 1 \end{pmatrix}, \quad V_{EL,OL} = V_{OR,ER}^T = \begin{pmatrix} 1+t & 1+t \\ 1 & 0 \\ 0 & 1 \end{pmatrix},$$

$$V_{EL,OR} = V_{OL,ER}^T = \begin{pmatrix} 1 & t \\ 0 & 1 \\ 0 & 0 \end{pmatrix}, \quad V_{OL,OL} = V_{OR,OR} = \begin{pmatrix} 1 & t \\ t & 1 \end{pmatrix} \quad \text{and}$$

$$V_{EL,ER} = \begin{pmatrix} 1+t & 1 & 0 \\ 1 & 0 & 0 \\ 0 & 0 & 0 \end{pmatrix}.$$

Using these matrices V and H we can then calculate $M = HV$ and

$(V(I - M)^{-1})_{\phi_L \phi_R}$. Together with Theorem 9 we obtain

$$F_{\mathbb{L}}(t) = \frac{1 + 2t - t^3 - t^4 + t^7}{(1 - t)^2(1 + t)^2(1 - t - t^2)} \tag{4.3}$$

as the generating function of SAWs on the ladder graph $\mathbb{L}$. Its radius of convergence R is the root of the denominator of $F_{\mathbb{L}}(t)$ having the smallest absolute value:

$$R = \frac{\sqrt{5} - 1}{2} \quad \text{and} \quad \mu = \frac{1}{R} = \frac{1 + \sqrt{5}}{2}.$$

5. Context-free languages

5.1. Grammars and formal languages

Definition 10. An *alphabet* Σ is a finite set of elements called letters. A word w is an element in some k-fold Cartesian product Σ^k and $|w| = k$ is called the *length* of w. We write $w = a_1 a_2 \cdots a_k$ to denote a *word* of length k and $\Sigma^* = \bigcup_{k \geq 0} \Sigma^k$ for the set of all words or strings of finite length over the alphabet Σ.

By defining a multiplication

$$\Sigma^* \times \Sigma^* \to \Sigma^*, (a_1 \cdots a_k, b_1 \ldots b_l) \mapsto a_1 \cdots a_k b_1 \cdots b_l$$

Σ^* becomes a monoid with neutral element ϵ, the empty word of length zero.

Definition 11. A *grammar* is a tuple $\mathcal{G} = (N, \Sigma, P, S)$ where

- N is a finite set of *non-terminal* symbols or *variables*,
- Σ is a finite set of *terminal* symbols (alphabet of the language),
- $P \subset (N \cup \Sigma)^* N (N \cup \Sigma)^* \times (N \cup \Sigma)^*$ is the set of *productions*,
- $S \in N$ is the *start symbol*.

We usually write $\lambda \to \rho \in P$ if $(\lambda, \rho) \in P$. $\lambda \to \rho_1 \mid \rho_2 \mid \cdots \mid \rho_k$ is a shorter way to describe k productions $\lambda \to \rho_1, \lambda \to \rho_2, \ldots, \lambda \to \rho_k$.

Define a binary relation $\Rightarrow$ on $(N \cup \Sigma)^*$: $x \Rightarrow y$ if and only if there are $u, v, p, q \in (N \cup \Sigma)^*$ such that $x = upv$, $y = uqv$ and $p \to q \in P$. We say that y derives from x in a single step.

The relation $\overset{*}{\Rightarrow}$ on $(N \cup \Sigma)^*$ is defined as the reflexive transitive closure of $\Rightarrow$, meaning that $x \overset{*}{\Rightarrow} y$ if and only if there is an integer $n \geq 0$ such that there are $v_0, v_1, \ldots, v_n \in (N \cup \Sigma)^*$ with $x = v_0 \Rightarrow v_1 \Rightarrow \cdots \Rightarrow v_n = y$. We say that y derives from x in n steps.

The language generated by the grammar $\mathcal{G}$ is the set $L(\mathcal{G}) = \{w \in \Sigma^* \mid S \overset{*}{\Rightarrow} w\}$.

Grammars can be classified in different ways. The most common one is the Chomsky hierarchy first described by Noam Chomsky in 1956.

© Springer Fachmedien Wiesbaden GmbH, part of Springer Nature 2018
C. Lindorfer, *The Language of Self-Avoiding Walks*, BestMasters,
https://doi.org/10.1007/978-3-658-24764-5_5

Definition 12. A grammar $\mathcal{G}$ and its generated language $L(\mathcal{G})$ are said to be of Type-i for $i \in \{0, 1, 2, 3\}$ if the following holds:

- Type-0 *(recursively enumerable):* No restrictions on $\lambda \to \rho \in P$.

- Type-1 *(context-sensitive or monotone):* All productions $\lambda \to \rho \in P$ with $(\lambda, \rho) \neq (S, \epsilon)$ must satisfy $|\lambda| \leq |\rho|$. If $S \to \epsilon \in P$, then S must not appear in the right-hand-side of any production.

- Type-2 *(context-free):* All productions $\lambda \to \rho \in P$ satisfy $\lambda \in N$.

- Type-3 *(regular):* All productions $\lambda \to \rho \in P$ satisfy $\lambda \in N$ and $\rho \in \Sigma^* N \cup \{\epsilon\}$.

Remark 7. It is known that Type-i languages form a subclass of Type-j languages for $i > j$ and that no two classes coincide.

Example 10. We give some standard examples for languages and the grammars $\mathcal{G} = (N, \Sigma, P, S)$ generating them. For our examples we use $\Sigma = \{a, b, c\}$ and $N = \{S, B, C\}$.

- $L_0 = \{a^k b^l c^m \mid k, l, m \geq 1\}$ is regular. A set of productions generating L_0 is

$$S \to aS \mid aB,$$
$$B \to bB \mid bC,$$
$$C \to cC \mid c.$$

- $L_1 = \{a^k b^k c^l \mid k, l \geq 1\}$ is context-free but not regular. A set of productions generating L_1 is

$$S \to BC,$$
$$B \to aBb \mid ab,$$
$$C \to cC \mid c.$$

- $L_2 = \{a^k b^k c^k \mid k \geq 1\}$ is context-sensitive but not context-free. A set of productions generating L_2 is

$$S \to aSBC \mid aBC,$$
$$CB \to BC,$$
$$aB \to ab,$$
$$bB \to bb,$$
$$bC \to bc,$$
$$cC \to cc.$$

We will now start working with context-free languages. Assume we have more than one non-terminal at some step of a derivation. As there is exactly one non-terminal at the left-hand side of each production (and nothing else), it clearly does not matter in which order productions are used. Therefore we will usually replace the leftmost variable first according to the following definition.

Definition 13. Let $\mathcal{G} = (N, \Sigma, P, S)$ be a context-free grammar generating the language $L(\mathcal{G})$. We call a step of derivation as described in Definition 11 *leftmost*, if the leftmost non-terminal is rewritten. We denote for a word $w \in \Sigma^*$ by $N(\mathcal{G}, w)$ the number of different derivation sequences of the form

$$S \Rightarrow \cdots \Rightarrow w$$

where each step of derivation is leftmost. We say that w is *generated unambiguously* by the grammar $\mathcal{G}$ if $N(\mathcal{G}, w) = 1$ and call $\mathcal{G}$ an *unambiguous grammar*, if every word $w \in L(\mathcal{G})$ is generated unambiguously by $\mathcal{G}$.

A method to study the structure of a context-free grammar is by studying its dependency-digraph introduced in the following definition taken from [3].

Definition 14. Let $\mathcal{G} = (N, \Sigma, P, S)$ be a context-free grammar. The *dependency-digraph* $\mathcal{D}(\mathcal{G}) = (N, E)$ is a directed graph which is allowed to have loops and has set of vertices N and oriented set of edges

$$E = \{(A, B) \in N \times N \mid \exists (A \to \rho) \in P \text{ such that } B \text{ occurs in } \rho\}.$$

In other words, an edge (A, B) appears in $\mathcal{D}(\mathcal{G})$ if and only if we can get to the non-terminal B from A in one step of derivation.

A standard method to prove that a given language L is not regular, meaning that there is no regular grammar producing L is the so-called Pumping Lemma for regular languages. The proof is very simple and uses the fact that there are only finitely many non-terminals in each grammar and therefore for long enough derivation chains there is a non-terminal appearing more than once.

Lemma 6. *For every regular language L over the alphabet Σ there is an integer $p \geq 1$, such that every word $w \in L$ with $|w| \geq p$ can be written in the form $w = xyz$ with strings $x, y, z \in \Sigma^*$ such that $|y| \geq 1, |xy| \leq p$ and $xy^i z \in L$ for every integer $i \geq 0$.*

5.2. Generating functions of context-free languages

Our next goal is to use a grammar $\mathcal{G}$ to get a formal power series which also generates the language $L(\mathcal{G})$ generated by $\mathcal{G}$. Chomsky and Schützenberger developed the following theory in [4].

Take any finite alphabet Σ and the monoid Σ^* defined in 10. Let r be a mapping which assigns to each word w in Σ^* an integer $\langle r, w \rangle$. We represent this map by a formal power series also denoted by r:

$$r = \sum_{w \in \Sigma^*} \langle r, w \rangle w.$$

We define the support of r as the set of strings with non-zero coefficients

$$Supp(r) = \{w \in \Sigma^* \mid \langle r, w \rangle \neq 0\}$$

and if for every $w \in \Sigma^*$ the coefficient $\langle r, w \rangle$ is either 0 or 1, we say that r is the characteristic formal power series of its support. For r and r' formal power series over the same alphabet Σ and n an integer define by

- nr the power series with coefficients $\langle nr, w \rangle = n \langle r, w \rangle$,
- $r + r'$ the power series with coefficients $\langle r + r', w \rangle = \langle r, w \rangle + \langle r', w \rangle$,
- rr' the power series with coefficients $\langle rr', w \rangle = \sum\limits_{\substack{w_1, w_2 \in \Sigma^* \\ w_1 w_2 = w}} \langle r, w_1 \rangle \langle r', w_2 \rangle$.

We call two formal power series r and r' equivalent mod degree n and write $r \equiv r'$ (mod deg n) if $\langle r, w \rangle = \langle r', w \rangle$ for every word w with $|w| \leq n$. Suppose now that we have an infinite sequence of formal power series $r_1, r_2, \ldots$ such that for all integers $n' \geq n \geq 1$ we have $r_n \equiv r_{n'}$ (mod deg n). In this case the limit r of the sequence $r_1, r_2, \ldots$ can be well defined as

$$r = \lim_{n \to \infty} \pi_n r_n$$

where for each n, $\pi_n r_n$ is the polynomial we get by replacing all coefficients $\langle r, w \rangle$ for $|w| \geq n$ by zero.

It is natural to associate with an unambiguous context-free grammar $\mathcal{G} = (N, \Sigma, P, S)$ the formal power series $r(\mathcal{G})$ having as coefficients

$$\langle r(\mathcal{G}), w \rangle = N(\mathcal{G}, w),$$

where $N(\mathcal{G}, w)$ is the degree of structural ambiguity of the word w defined in Definition 13. We call $r(\mathcal{G})$ the generating function of $L(\mathcal{G})$. Then the support of $r(\mathcal{G})$ is exactly the language generated by $\mathcal{G}$.

Suppose $N = \{S = V_1, \ldots, V_n\}$ are the non-terminals of the context-free grammar $\mathcal{G} = (N, \Sigma, P, S)$ and let P be given by

$$V_i \to \rho_{i,1} \mid \rho_{i,2} \mid \cdots \mid \rho_{i,m_i} \text{ for all } 1 \leq i \leq n.$$

We will assume that the grammar $\mathcal{G}$ contains no productions of the form

$$V_i \to \epsilon,$$
$$V_i \to V_j$$

and that for every i there must be (non-empty) words in the language of strings derivable from V_i. It is not hard to show that for every context-free grammar $\mathcal{G}$ not satisfying these assumptions and generating the language $L(\mathcal{G})$, there is a second context-free grammar $\mathcal{G}'$ which does satisfy them and also generates $L(\mathcal{G})$ (or $L(\mathcal{G}) \setminus \{\epsilon\}$, if the empty word ϵ is in $L(\mathcal{G})$). We associate for every i with V_i the polynomial expression

$$\sigma_i = \rho_{i,1} + \rho_{i,2} + \cdots + \rho_{i,m_i}$$

and with the grammar $\mathcal{G}$ the set of equations

$$V_1 = \sigma_1; \ldots; V_n = \sigma_n. \tag{5.1}$$

For every i we can use the equation $V_i = \sigma_i$ in (5.1) for defining a mapping ψ_i taking an n-tuple $(r_1, \ldots, r_n)$ of formal power series to the power series obtained by replacing all variables V_j appearing in σ_i by r_j. Combining all of these mappings we get a mapping ψ defined by

$$\psi(r_1, \ldots, r_n) = (\psi_1(r_1, \ldots, r_n), \ldots, \psi_n(r_1, \ldots, r_n)). \tag{5.2}$$

Consider now the infinite sequence of n-tuples $(r_1^{(k)}, \ldots, r_n^{(k)})_{k \geq 0}$ of power series defined iteratively by

$$\begin{aligned} r_i^{(0)} &= 0 \quad \text{for } 1 \leq i \leq n, \\ r_i^{(j)} &= \psi_i(r_1^{(j-1)}, \ldots, r_n^{(j-1)}) \quad \text{for } 1 \leq i \leq n, \, j \geq 1. \end{aligned} \tag{5.3}$$

Each $r_i^{(j)}$ has only finitely many non-zero coefficients. Furthermore using our assumptions on the grammar $\mathcal{G}$ it can be shown that

$$r_i^{(j)} \equiv r_i^{(j')} \pmod{\deg j} \quad \text{for all } 0 < j < j', \, 1 \leq i \leq n.$$

Therefore the limit $r_i^{(\infty)}$ of the infinite sequence $(r_i^{(j)})_{j \geq 0}$ is well defined for each $1 \leq i \leq n$. It is the only n-tuple within our framework to satisfy the equations (5.1) given by our grammar $\mathcal{G}$. In particular $r_1^{(\infty)}$ which is the series corresponding to our start symbol S is the generating function of $L(\mathcal{G})$, which we called $r(\mathcal{G})$ above. The following example gives an idea of the process described above.

Example 11. Consider the context-free grammar $\mathcal{G} = (N, \Sigma, P, S)$ with $N = \{S\}$, $\Sigma = \{a, b\}$ and productions

$$S \to bSS \mid a.$$

The equation described in (5.1) reads as

$$S = a + bSS. \tag{5.4}$$

This yields the mapping ψ from (5.2) defined on a power series r by

$$\psi(r) = a + brr.$$

Iteration as shown in (5.3) results in the sequence

$$
\begin{aligned}
r^{(0)} &= 0, \\
r^{(1)} &= a + br_0 r_0 = a, \\
r^{(2)} &= a + br_1 r_1 = a + baa, \\
r^{(3)} &= a + br_2 r_2 = a + b(a + baa)(a + baa) \\
&= a + baa + babaa + bbaaa + bbaabaa.
\end{aligned}
$$

$$\vdots \qquad\qquad \vdots$$

Clearly for $0 < j < j'$ we have $r_{(j)} \equiv r_{(j')} \pmod{\deg j}$ and therefore the limit $r^{(\infty)}$ is well defined and it is the characteristic generating function of the language $L(\mathcal{G})$.

We will now get back to classical (commutative) generating functions. For ρ and ρ' in $(N \cup \Sigma)^*$ we write $\varphi\rho = \varphi\rho'$ if they contain exactly the same number of each letter and non-terminal. Clearly this map φ extends to a mapping from our non-commutative formal power series onto the ring of ordinary commutative formal power series with integral coefficients. For a grammar $\mathcal{G}$ and its generating function $r(\mathcal{G})$ we call $\varphi r(\mathcal{G})$ the ordinary generating function of $\mathcal{G}$. Clearly it is the solution of the commutative version of the system of equations (5.1). This can be utilized to obtain $\varphi r(\mathcal{G})$ in a simple way.

Example 12. Consider again the grammar $\mathcal{G}$ of Example 11. From (5.4) we get the equation

$$\varphi r(\mathcal{G}) = \varphi a + \varphi b (\varphi r(\mathcal{G}))^2$$

admitting the two solutions

$$\varphi r(\mathcal{G}) = \frac{1 \pm \sqrt{1 - 4\varphi a \varphi b}}{2\varphi b}.$$

We want our ordinary generating function to have only positive coefficients, so we take the solution with minus. Using the binomial formula and simplifying gives the series

$$\varphi r(\mathcal{G}) = \sum_{n \geq 0} \frac{1}{n+1} \binom{2n}{n} (\varphi a)^{n+1} (\varphi b)^n.$$

This shows us that the number of words in $L(\mathcal{G})$ containing exactly $(n+1)$ a's and n b's is the n-th Catalan number.

In this context the following class of grammars lying between regular and context-free languages is of remarkable interest.

Definition 15. We call a context free grammar $\mathcal{G} = (N, \Sigma, P, S)$ *linear*, if all productions contain at most one non-terminal on their right hand side. Productions of this type are also called *linear*.

Now in the case of $\mathcal{G}$ being a linear grammar with alphabet $\Sigma = \{a_1, \ldots a_k\}$, the commutative version of the system of equations (5.1) is a linear system of equations with coefficients being polynomials in the variables $a_1, \ldots a_k$. Clearly this system can be solved and gives rise to a solution in the field of rational functions in the commutative variables $a_1, \ldots a_k$. Therefore the generating function $r(\mathcal{G})$ can be written as a rational function.

To get this property for even more context-free grammars we use the dependency-digraph from Definition 14. Let $\mathcal{D}(\mathcal{G}) = (N, E)$ be the dependency-digraph of the context-free grammar $\mathcal{G} = (N, \Sigma, P, S)$. Assume there is a subset C of N such that there are no edges going from C to $N \setminus C$ in $\mathcal{D}(\mathcal{G})$ and such that for all $V \in C$ all productions $V \to \rho \in P$ are linear. Consider the subsystem of equations of (5.1) we obtain by only taking the lines corresponding to the non-terminals in C. By the definition of the dependency-digraph no non-terminals of $N \setminus C$ can appear in this subsystem. So we can solve the subsystem as mentioned above and get rational ordinary generating functions for all non-terminals in C. Now if for a non-linear production $V \to \rho \in P$, all non-terminals appearing in ρ are already known to have rational generating functions, clearly also the generating function corresponding to V is rational as the sum over products of rational functions. This Observation leads to the following definition:

Definition 16. An *ordered cycle* of length $n \geq 1$ in a directed graph which is allowed to have loops is a sequence of edges $(e_1, \ldots, e_n)$ such that $e_i^+ = e_{i+1}^-$ for all $1 \leq i \leq n-1$ and $e_n^+ = e_1^-$.
In the dependency-digraph $\mathcal{D}(\mathcal{G})$ of a given context-free grammar $\mathcal{G} = (N, \Sigma, P, S)$ we call an edge (A, B) *non-linear* if there is a non-linear production $A \to \rho \in P$ such that B appears in ρ, else we call the edge *linear*. We call the context-free Grammar $\mathcal{G}$ *ultimately linear*, if in the dependency-digraph $\mathcal{D}(\mathcal{G})$ no non-linear edge is contained in an ordered cycle.

Let an ultimately linear grammar $\mathcal{G} = (N, \Sigma, P, S)$ be given. We denote for a non-linear edge (A, B) by $C(B)$ the set of all non-terminals which can be reached from B by following a (directed) walk in $\mathcal{D}(\mathcal{G})$. Clearly there can be no edge going from $C(B)$ to $N \setminus C(B)$. If all non-terminals in $C(B)$ have rational ordinary generating functions, this is also true for A. Iteratively using this argument starting at an edge (A, B) such that no two non-terminals in $C(B)$ are connected by a non-linear edge we get that the generating function corresponding to the language $\mathcal{G}$ is also rational. Such an edge must always exist as the graph $\mathcal{D}(\mathcal{G})$ is finite and there are no cycles containing non-linear edges. An ultimately linear grammar can be found in Example 15.

6. The language of self-avoiding walks

6.1. Edge-labelled graphs

Definition 17. A directed edge-labelled graph G is a tuple $G = (V, E, \Sigma, l)$ where

- V and E are set of vertices and edges of a directed graph,
- Σ is the *label alphabet*,
- $l : E \to \Sigma$ is the *label function*.

We call such a graph G

- *fully labelled* if for all $u \in V$ and $a \in \Sigma$ there is a vertex $v \in V$ such that $(u, v) \in E$ and $l((u, v)) = a$,
- *deterministic* if any two edges e_1, e_2 starting at the same vertex have different labels $l(e_1) \neq l(e_2)$,
- *symmetric* if for every $a \in \Sigma$ there is a letter $b \in \Sigma$ such that $(u, v) \in E$ with $l(u, v) = a$ if and only if $(v, u) \in E$ with $l(v, u) = b$. In this case we call b the *inverse* of a.

For a given set of walks on a labelled graph we are now able to define the language corresponding to it.

Definition 18. Let $G = (V, E, \Sigma, l)$ be a directed edge-labelled graph. A *walk* of length n in G connecting two vertices u and v is a sequence of edges $\pi = (e_1, e_2 \ldots, e_n)$ such that $e_1^- = u$, $e_n^+ = v$ and $e_i^+ = e_{i+1}^-$ for $i = 1, \ldots, n - 1$.
We extend the map l to walks $\pi = (e_1, e_2 \ldots, e_n)$ by

$$l(\pi) = l(e_1)l(e_2) \ldots l(e_n) \in \Sigma^*.$$

So for every walk π we get a word in Σ^* by reading the letters of the edges contained in the walk.
Let Π be a set of walks in G. Then we call $L(\Pi) = \{l(\pi) \mid \pi \in \Pi\} \subset \Sigma^*$ the *language of* Π.

Remark 8. Let $G = (V, E, \Sigma, l)$ be a directed edge-labelled graph which is deterministic and let $v \in V$ be given. For any set Π of paths starting

© Springer Fachmedien Wiesbaden GmbH, part of Springer Nature 2018
C. Lindorfer, *The Language of Self-Avoiding Walks*, BestMasters,
https://doi.org/10.1007/978-3-658-24764-5_6

at v, the extension of the label function l is a natural bijection between Π and $L(\Pi)$. This means that for any given word w in $L(\Pi)$ we can get the unique path π in Π with $w = l(\pi)$ by following the edges labelled with the letters of w.

We will now extend the notions of transitivity and quasi-transitivity onto directed edge-labelled graphs.

Definition 19. The *automorphism group* of a directed edge-labelled graph $G = (V, E, \Sigma, l)$ denoted by $AUT(G)$ is the group of all permutations $\sigma : V \to V$ such that for all $u, v \in V$ and $a \in \Sigma$ we have: $(u, v) \in E$ with $l((u, v)) = a$ if and only if $(\sigma(u), \sigma(v)) \in E$ with $l((\sigma(u), \sigma(v))) = a$.
A subgroup $\Gamma \leq AUT(G)$ is said to act *transitively* on G if, for any $u, v \in V$, there exists $\gamma \in \Gamma$ with $\gamma u = v$. It is said to act *quasi-transitively* if there exists a finite set $W \subset V$ such that for any $u \in V$ there exist $v \in W$ and $\gamma \in \Gamma$ with $\gamma u = v$.
The directed edge-labelled graph G is called *transitive* (respectively *quasi-transitive*) if its automorphism group $AUT(G)$ acts transitively (respectively quasi-transitively) on G.

We note that when taking a undirected transitive or quasi-transitive graph G and adding any edge labels, we do not have to end up with a transitive or quasi-transitive directed edge-labelled graph. We also want to get a deterministic graph as mentioned in Remark 8. So it is important to add the labels in a convenient way. The following example shows two good ways to add labels to the ladder graph $\mathbb{L}$.

Example 13. Consider the ladder graph $\mathbb{L} = \mathbb{Z} \times \{0, 1\}$. We will first label the (directed) edges with elements of the alphabet $\Sigma = \{s, a, b\}$ to get the labelled directed graph $\mathbb{L}_1$ in the following way also shown in Figure 6.1:

$$l(e) = \begin{cases} s & \text{if} \quad e = ((x, 0), (x, 1)) \text{ or } e = ((x, 1), (x, 0)) \text{ for some } x \in \mathbb{Z}, \\ a & \text{if} \quad e = ((x, y), (x+1, y)) \text{ or} \\ & \qquad e = ((x+1, y), (x, y)) \text{ for some even } x \in \mathbb{Z}, \, y \in \{0, 1\}, \\ b & \text{if} \quad e = ((x, y), (x+1, y)) \text{ or} \\ & \qquad e = ((x+1, y), (x, y)) \text{ for some odd } x \in \mathbb{Z}, \, y \in \{0, 1\}. \end{cases}$$

The labelled graph $\mathbb{L}_1$ is transitive: The automorphism group $AUT(\mathbb{L}_1)$ contains the reflection $(x, y) \mapsto (1 - x, y)$, the horizontal translations $(x, y) \mapsto (x + 2k, y)$ for every integer k and the reflection $(x, y) \mapsto (x, 1 - y)$ and these automorphisms can be concatenated to map $(0,0)$ onto any given vertex. Also $\mathbb{L}_1$ is fully labelled, deterministic and symmetric. Moreover, each label is inverse to itself, so we could also draw this graph as undirected graph and only add one label per edge.

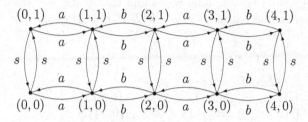

Figure 6.1: Labelled directed ladder $\mathbb{L}_1$.

The more "natural" way to label the edges of the directed version of $\mathbb{L}$ is to label the edges using their direction as done by Zeilberger in [22]. We will use the alphabet $\Sigma = \{r, l, u, d\}$ in the following way also shown in Figure 6.2 to get the labelled directed graph $\mathbb{L}_2$:

$$
l(e) = \begin{cases}
r & \text{if } e = ((x, y), (x + 1, y)) \text{ for some } x \in \mathbb{Z}, y \in \{0, 1\}, \\
l & \text{if } e = ((x + 1, y), (x, y)) \text{ for some } x \in \mathbb{Z}, y \in \{0, 1\}, \\
u & \text{if } e = ((x, 0), (x, 1)) \text{ for some } x \in \mathbb{Z}, \\
d & \text{if } e = ((x, 1), (x, 0)) \text{ for some } x \in \mathbb{Z}.
\end{cases}
$$

It is easy to see that $\mathbb{L}_2$ is not transitive: Vertices on the line $\mathbb{Z} \times \{0\}$ have an outgoing edge labelled by u, while vertices on line $\mathbb{Z} \times \{1\}$ do not have such an edge. Therefore we cannot have an element in $AUT(\mathbb{L}_2)$ mapping the vertex $(0, 0)$ to the vertex $(0, 1)$. However $\mathbb{L}_2$ is quasi-transitive as all horizontal translations mapping (x, y) to $(x + k, y)$ for some k in $\mathbb{Z}$ are in $AUT(\mathbb{L}_2)$. Clearly $\mathbb{L}_2$ is deterministic and symmetric (the inverse of u being d and the inverse of l being r), but not fully labelled.

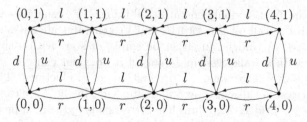

Figure 6.2: Labelled directed ladder $\mathbb{L}_2$.

6.2. The language of SAWs and bridges on the ladder

In this section our goal is to calculate the characteristic generating function of the language of self-avoiding walks and the language of bridges on the edge-labelled ladder graph. Let $\mathbb{L} = (V, E, \Sigma, l)$ be the edge-labelled ladder graph defined by

- $V = \mathbb{Z} \times \{0, 1\}$.
- $E = E_r \cup E_l \cup E_s$ where
  $E_r = \{((x, y), (x + 1, y)) \mid x \in \mathbb{Z}, y \in \{0, 1\}\}$,
  $E_l = \{((x + 1, y), (x, y)) \mid x \in \mathbb{Z}, y \in \{0, 1\}\}$,
  $E_s = \{((x + 1, y), (x, y)) \mid x \in \mathbb{Z}, y \in \{0, 1\}\}$.
- $\Sigma = \{r, l, s\}$.
- $l : E \to \Sigma$ defined by $l(e) = a$ if and only if $e \in E_a$ for $a \in \Sigma$.

As in Example 4 we take the pair $(h, \mathcal{H})$ as unimodular graph height function, where h is the map $V \to \mathbb{Z}$, $(x, y) \mapsto x$ and $\mathcal{H}$ is the set of all horizontal translations $(x, y) \mapsto (x + k, y)$ for some $k \in \mathbb{Z}$.

Example 14. We start by constructing a grammar generating L_B, the language of all bridges starting at the origin $(0, 0)$. As already mentioned in Example 4, every bridge on $\mathbb{L}$ is of the form L^*I where L denotes again a sub-walk of the form $r^i s$ for some positive integer i and I stands for a sub-walk of the form r^j for some non-negative integer j. Here L^* denotes a sequence of length ≥ 0 of walks of type L. We can translate this into the unambiguous regular grammar $\mathcal{G}_B = (N_B, \Sigma, P_B, S)$, where Σ comes from the graph $\mathbb{L}$, $N_B = \{S, L, I\}$ and P_B is the following set of productions generating L_B:

$$
\begin{aligned}
S &\to V_L \mid V_I \mid \epsilon, \\
V_L &\to rV_L \mid rsS, \\
V_I &\to rV_I \mid r.
\end{aligned}
\tag{6.1}
$$

Here S is the start symbol generating all bridges, V_L generates all bridges starting with a walk of the form L and V_I generates all walks of form I of length ≥ 1. A bridge can either be empty or start with a sub-walk of form L or with a walk of form I (only if it does not contain a sub-walk of form L). This corresponds to the first line of productions. The second line produces sub-walks of type L and after finishing an L we can continue again with any bridge. The third line corresponds to the final piece of type I of a bridge. Note that the Grammar $\mathcal{G}_B$ is not only context-free, but even regular. We compute the generating function of L_B.

As described in Chapter 5 the equations for the ordinary generating functions F_S, F_L and F_I corresponding to our non-terminals S, V_L and V_I are obtained from the productions of $\mathcal{G}_B$ in (6.1):

$$F_S = F_L + F_I + 1, \tag{6.2}$$
$$F_L = rF_L + rsF_S, \tag{6.3}$$
$$F_I = rF_I + r. \tag{6.4}$$

Keep in mind that we are working in the non-commutative setting. However, from (6.4) we have

$$F_I = (1 - r)^{-1} r. \tag{6.5}$$

From (6.3) by using (6.2) and (6.5) we get

$$F_L = rF_L + rs(F_L + (1 - r)^{-1} r + 1).$$

Solving for F_L yields

$$F_L = (1 - r - rs)^{-1} rs (1 - r)^{-1}. \tag{6.6}$$

Plugging (6.5) and (6.6) into (6.2) and simplifying provides

$$F_S = (1 - r - rs)^{-1}$$

which is the desired characteristic generating function of the language L_B. We can now translate it back to achieve

$$F_S = \sum_{n \geq 0} (r + rs)^n = 1 + r + rs + r^2 + r^2 s + rsr + rsrs + \dots .$$

Every term in this series represents a word in the language L_B and therefore a bridge on $\mathbb{L}$. By replacing r and s by the new variable t we get again the generating function of bridges on the unlabelled graph $\mathbb{L}$ we already derived in Example 4.

In the next step we want to find a grammar for the language L_W of all SAWs on the labelled ladder graph $\mathbb{L}$ starting at the vertex $(0,0)$.

Remark 9. We have already seen that the language of bridges L_B is regular. This is not true for L_W. We can use the Pumping Lemma for regular languages to prove this:
Take any integer $p \geq 1$ and consider the walk $w = r^p s l^{p+1} s \in L_W$. Then for every decomposition $w = xyz$ with $|y| \geq 1$ and $|xy| \leq p$ we have that y is of the form r^k for some $k \geq 1$. But then $xy^2 z = r^{p+k} s l^{p+1} s$ corresponds to a path containing its end-vertex twice and is therefore not in L_W. So L_W does not satisfy the statement from Lemma 6 and thus cannot be regular.

Example 15. It is convenient to use the following notation for sub-walks of certain types:

- U_r is a walk of the form $r^i s l^i$ for some $i \geq 1$,
- L_r is a walk of the form $r^i s$ for some $i \geq 1$,
- I_r is a walk of the form r^i for some $i \geq 1$.

We define U_l, L_l and I_l as U_r, L_r and I_r, but by replacing every l by r and vice versa. We give an unambiguous context-free grammar $\mathcal{G}_W = (N_W, \Sigma, P_W, S)$ generating the language L_W: Our set of non-terminals is

$$N_W = \{S, V_A^{(r)}, V_B^{(r)}, V_L^{(r)}, V_I^{(r)}, V_U^{(r)}, V_A^{(l)}, V_B^{(l)}, V_L^{(l)}, V_I^{(l)}, V_U^{(l)}\}$$

and the set of productions P_W is given by the following productions together with the productions we get when replacing every r by l and vice versa in every production but the first one:

$$
\begin{aligned}
S \quad &\to V_A^{(r)} \mid V_A^{(l)} \mid V_B^{(r)} \mid V_B^{(l)} \mid sV_B^{(r)} \mid sV_B^{(l)} \mid s \mid \epsilon, \\
V_A^{(r)} &\to lV_U^{(l)} rV_B^{(r)} \mid lsrV_B^{(r)} \mid lV_U^{(l)} r \mid lsr, \\
V_B^{(r)} &\to V_L^{(r)} \mid V_I^{(r)}, \\
V_L^{(r)} &\to rV_L^{(r)} \mid rsV_L^{(r)} \mid rsV_I^{(r)} \mid rs, \\
V_I^{(r)} &\to rV_I^{(r)} \mid rV_U^{(r)} \mid r, \\
V_U^{(r)} &\to rV_U^{(r)} l \mid rsl.
\end{aligned}
\tag{6.7}
$$

The non-terminals defined above generate the language of all SAWs having the following properties:

- S: All walks.
- $V_A^{(r)}$: Walks starting with U_l.
- $V_B^{(r)}$: Walks starting with r, not containing the vertex above/below of the start vertex.
- $V_L^{(r)}$: Walks starting with L_r.
- $V_I^{(r)}$: Walks starting with I_r.
- $V_U^{(r)}$: Walks of the form U_r.

We note that because of the rule $V_A^{(r)} \to V_U^{(l)} rV_B^{(r)}$ the grammar is not linear. So we can not directly conclude that the ordinary generating function of L_W can be written as a rational function. This is where the dependency-digraph $\mathcal{D}(\mathcal{G}_W)$ shown in Figure 6.3 comes into play. The only non-linear edges from Definition 16 in are $(V_A^{(r)}, V_B^{(r)}), (V_A^{(r)}, V_U^{(l)}), (V_A^{(l)}, V_B^{(l)})$ and $(V_A^{(l)}, V_U^{(r)})$. There are no edges going from $C^{(i)} = \{V_B^{(i)}, V_L^{(i)}, V_I^{(i)}, V_U^{(i)}\}$ to $N \setminus C^{(i)}$ for $i \in \{l, r\}$, so our language is ultimately linear. Therefore all appearing ordinary generating functions are rational.

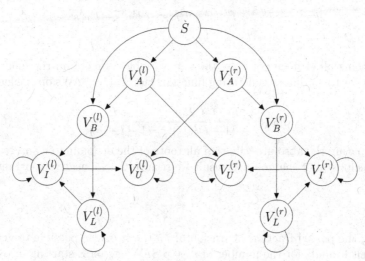

Figure 6.3: Dependency-digraph $\mathcal{D}(\mathcal{G}_W)$.

Using the productions in (6.7) we get the following equations for the ordinary generating functions corresponding to the non-terminals of $\mathcal{G}_W$:

$$F_S = F_A^{(r)} + F_A^{(l)} + F_B^{(r)} + F_B^{(l)} + sF_B^{(r)} + sF_B^{(l)} + s + 1, \qquad (6.8)$$

$$F_A^{(r)} = lF_U^{(l)}rF_B^{(r)} + lsrF_B^{(r)} + lF_U^{(l)}r + lsr, \qquad (6.9)$$

$$F_B^{(r)} = F_L^{(r)} + F_I^{(r)}, \qquad (6.10)$$

$$F_L^{(r)} = rF_L^{(r)} + rsF_L^{(r)} + rsF_I^{(r)} + rs, \qquad (6.11)$$

$$F_I^{(r)} = rF_I^{(r)} + rF_U^{(r)} + r, \qquad (6.12)$$

$$F_U^{(r)} = rF_U^{(r)}l + rsl. \qquad (6.13)$$

Again the equations for generating functions with superscript (l) are defined similar by exchanging l and r. As mentioned above the subsystem of equations (6.10) - (6.13) is linear and can be solved without considering the other equations. The following result is the ordinary (commutative) generating function of our language L_W and was obtained by using Sage for the calculations:

$$F_S = \frac{N}{D}$$

where

$$N = 1 + s - 3lr - 3lrs - 3lrs^2 + 3l^2r^2 - lrs^3 + 3l^2r^2s +$$
$$2l^2r^2s^2 - l^3r^3 + l^2r^3s^2 + l^3r^2s^2 - l^3r^3s - l^3r^3s^2 - l^3r^3s^3,$$
$$D = (1 - lr)^2(1 - l - ls)(1 - r - rs).$$

Replacing all elements of Σ by the new variable t we obtain the following result for the ordinary generating function $F_W(t)$ of all SAWs on the ladder graph $\mathbb{L}$:

$$F_W(t) = \frac{1 + 2t - t^3 - t^4 + t^7}{(1 - t)^2(1 + t)^2(1 - t - t^2)}. \tag{6.14}$$

Calculating the absolute values of all roots of the denominator and taking the reciprocal of the minimal result we receive the connective constant

$$\mu(\mathbb{L}) = \frac{1 + \sqrt{5}}{2}.$$

Using the partial fraction expansion of $F_W(t)$, it is also possible to get an explicit formula for the number of n-step SAWs σ_n on $\mathbb{L}$ starting in $(0,0)$. This was done by Zeilberger in [22] and results in $\sigma_0 = 1$, $\sigma_1 = 3$ and

$$\sigma_n = 8F_n - \frac{n}{2}(1 + (-1)^n) - 2(1 - (-1)^n) \quad \text{for all } n \geq 2$$

where F_n denotes the n-th Fibonacci Number.

6.3. The language of SAWs on the k-ladder-tree

Consider for a fixed integer $k \geq 2$ the k-regular tree $\mathbb{T}_k = (V(\mathbb{T}_k), E(\mathbb{T}_k))$, where the edges are labelled with $a_1, a_2, \ldots, a_k$, such that no two edges having the same label start at the same vertex and each pair of edges corresponding to an undirected edge has the same label. Clearly the labelled $\mathbb{T}_k$ is a transitive graph and by Example 1 its connective constant is $\mu(\mathbb{T}_k) = k - 1$.

We take two copies of $\mathbb{T}_k$ and connect two vertices of different copies if they correspond to the same vertex in $\mathbb{T}_k$ and label these new edges with s. We denote the result shown in Figure 6.4 by $\mathbb{LT}_k$ and call it k-ladder-tree.

A formal definition of the k-ladder-tree is $\mathbb{LT}_k = (V, E, \Sigma, l)$, where the set of vertices is $V = V(\mathbb{T}_k) \times \{0, 1\}$, the set of edges is $E(\mathbb{LT}_k) = E_T \cup E_L$,

$$E_T = \{((u, x), (v, x)) \mid (u, v) \in E(\mathbb{T}_k), x \in \{0, 1\}\},$$
$$E_L = \{((v, x), (v, 1 - x)) \mid v \in V(\mathbb{T}_k), x \in \{0, 1\}\}$$

and the label alphabet is $\Sigma = \{a_1, a_2, \ldots, a_k, s\}$. Edges in E_T are called tree-edges and inherit their labels from $\mathbb{T}_k$ and edges in E_L are link-edges labelled by s.

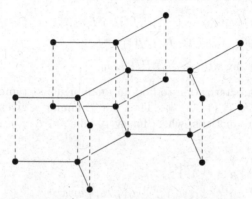

Figure 6.4: 3-ladder-tree, dashed edges are link-edges.

The resulting directed labelled tree $\mathbb{LT}_k$ clearly is deterministic and symmetric, each label being self-inverse and it is also transitive.

We fix a vertex $(v, 0)$ of $\mathbb{LT}_k$ and want to find the language L_W of SAWs starting at $(v, 0)$. Start by trying to find a characterization of SAWs. It is again convenient to use the following notation similar to the one in Example 15:

- U is a walk of the form $b_1 \ldots b_i s b_i \ldots b_1$,

- L is a walk of the form $b_1 \ldots b_i s$,

- I is a walk of the form $b_1 \ldots b_i$,

where in every case $i \geq 1$, $b_j \in \Sigma \setminus \{s\}$ for all $1 \leq j \leq i$ and $b_j \neq b_{j+1}$ for all $1 \leq j \leq i - 1$.

Consider for a SAW π on $\mathbb{LT}_k$ the "projection" π' of π onto the tree $\mathbb{T}_k$ obtained by removing all link-edges from the path. Clearly π' does not have to be self-avoiding, but every vertex can appear at most twice. Whenever a vertex v appears twice in this walk, the sub-walk between the first and second appearance of v has to be of the form $b_1 \ldots b_i b_i \ldots b_1$ for some $i \geq 1$, otherwise a vertex would appear more than twice. We denote by U' such paths from v to v. Between two walks of type U' there has to be a walk of type I. So all walks π' not containing the start vertex twice have to be of the form $(IU')^* I_0$, where I_0 denotes a walk of type I or the empty path. Going back to $\mathbb{LT}_k$, every walk starting at $(v, 0)$ and not containing $(v, 1)$ has to be of the form $(L^* IU)^* L^* I_0$, as a walk of type U' translates back to U and a walk of type I translates back to $L^* I$. This gives us the following characterization for all SAWs on $\mathbb{LT}_k$:

- SAWs not containing $(v, 1)$: $(L^* IU)^* L^* I_0$,

- paths starting with U: $U(L^* IU)^* L^* I_0$,

- paths starting with s: $s(L^* IU)^* L^* I_0$.

We use this characterization to find an unambiguous grammar $\mathcal{G}_W$ generating L_W. Let $K = \{1, \ldots, k\}$. Then our grammar is given by the set of the following productions, where for every rule $i \in K, j, j' \in K \setminus \{i\}$ and $l \in K \setminus \{i, j\}$:

$$
\begin{aligned}
S \quad &\to \quad V_A^{(i)} \mid V_B^{(i)} \mid sV_B^{(i)} \mid s \mid \epsilon, \\
V_A^{(i)} \quad &\to \quad a_i V_U^{(j)} a_i V_B^{(j')} \mid a_i s a_i V_B^{(j)} \mid a_i V_U^{(j)} a_i \mid a_i s a_i, \\
V_B^{(i)} \quad &\to \quad V_L^{(i)} \mid V_I^{(i)} \mid V_E^{(i)} \mid V_F^{(i)}, \\
V_L^{(i)} \quad &\to \quad a_i V_L^{(j)} \mid a_i s V_L^{(j)} \mid a_i s V_I^{(j)}, \\
V_I^{(i)} \quad &\to \quad a_i V_I^{(j)} \mid a_i V_C^{(j,i)}, \\
V_C^{(i,j)} \quad &\to \quad a_i V_U^{(j)} a_i V_L^{(l)} \mid a_i V_U^{(j)} a_i V_I^{(l)} \mid a_i V_U^{(j)} a_i V_E^{(l)} \mid a_i V_U^{(j)} a_i V_F^{(l)} \mid a_i V_U^{(j)} a_i \mid \\
&\qquad a_i s a_i V_L^{(l)} \mid a_i s a_i V_I^{(l)} \mid a_i s a_i V_E^{(l)} \mid a_i s a_i V_F^{(l)} \mid a_i s a_i, \\
V_U^{(i)} \quad &\to \quad a_i V_U^{(j)} a_i \mid a_i s a_i, \\
V_E^{(i)} \quad &\to \quad a_i V_E^{(j)} \mid a_i s V_E^{(j)} \mid a_i s V_F^{(j)} \mid a_i s, \\
V_F^{(i)} \quad &\to \quad a_i V_F^{(r)} \mid a_i.
\end{aligned}
$$

$$(6.15)$$

The following list shows the properties of the walks generated by each non-terminal

- S: Start symbol, generates all SAWs starting at v_0.

- $V_A^{(i)}$: Walks of the form $U(L^* IU)^* L^* I_0$, first step a_i.

- $V_B^{(i)}$: Walks of the form $(L^* IU)^* L^* I_0$, first step a_i.

- $V_L^{(i)}$: Walks of the form $(LL^*IU)(L^*IU)^*L^*I_0$, first step a_i.

- $V_I^{(i)}$: Walks of the form $(IU)(L^*IU)^*L^*I_0$, first step a_i.

- $V_C^{(i,j)}$: Walks of the form $U(L^*IU)^*L^*I_0$, first step a_i, first step after initial U not a_j.

- $V_U^{(i)}$: Walks of the form U, first step a_i.

- $V_E^{(i)}$: Walks of the form LL^*I_0, first step a_i.

- $V_F^{(i)}$: Walks of the form I, first step a_i.

Observe that the grammar $\mathcal{G}_W$ is not linear. By using the dependency-digraph in the same way as in Example 15 it can be seen that it is ultimately linear and that the ordinary generating function of our language is a rational function. This was not clear up to this point, as $\mathbb{LT}_k$ is not a one-dimensional lattice.

The number of non-terminals of our language depends on k, so we do not directly solve the system of equations achieved from (6.15). To reduce the number of variables, we use the following idea: After replacing every letter of Σ by the new letter t, by symmetry of $\mathbb{LT}_k$ we clearly obtain the same equation for $V_A^{(i)}$ and $V_A^{(j)}$ for $i, j \in K$. So we can associate the ordinary generating function F_A in the variable t with the set of our non-terminals $V_A^{(i)}$ for all $i \in K$. Doing the same for all non-terminals (6.15) translates into the following system of equations still containing the parameter k:

$$
\begin{aligned}
F_S &= kF_A + kF_B + ktF_B + t + 1, \\
F_A &= (k-1)^2t^2F_UF_B + (k-1)t^3F_B + t^2F_U + t^3, \\
F_B &= F_L + F_I + F_E + F_F, \\
F_L &= (k-1)tF_L + (k-1)t^2F_L + (k-1)t^2F_I, \\
F_I &= (k-1)tF_I + (k-1)tF_C, \\
F_C &= (k-1)(k-2)t^2F_U(F_L + F_I + F_E + F_F) + (k-1)t^2F_U + \\
&\quad (k-2)t^3(F_L + F_I + F_E + F_F) + t^3, \\
F_U &= (k-1)t^2F_U + t^3, \\
F_E &= (k-1)tF_E + (k-1)t^2F_E + (k-1)t^2F_F + t^2, \\
F_F &= (k-1)tF_F + t.
\end{aligned}
$$

Again we used Sage to solve the above system of equations and obtain the ordinary generating function

$$F_S = \frac{N}{D}$$ (6.16)

where

$$N = 1 + 2t + (4 - 2k)t^2 + (5 - 3k)t^3 + (3 - 4k + k^2)t^4 +$$
$$(2 - 3k + k^2)t^5 - (1 - k)t^7,$$
$$D = (1 + (1 - k)t + (2 - 2k)t^2 + (1 - 2k + k^2)t^3 - (1 - k)t^4)$$
$$(1 + (1 - k)t^2).$$

Note that the variable t corresponds to a single step of any form. So the result for F_S is the generating function of SAWs we introduced in Definition 4. Using this generating function we can calculate the connective constants $\mu(\mathbb{LT}_k)$ for given values of k. The results for some values of k are contained in Table 6.1.

k	$\mu(\mathbb{LT}_k)$	k	$\mu(\mathbb{LT}_k)$
2	1.618034	10	9.981284
3	2.825955	20	19.995146
4	3.896361	50	49.999208
5	4.930990	100	99.999801
6	5.950746	1000	999.999998

Table 6.1: Connective constants of $\mu(\mathbb{LT}_k)$ (rounded values).

Clearly the 2-regular tree $\mathbb{T}_2$ is the line graph $\mathbb{Z}$, hence $\mathbb{LT}_2$ is the ladder graph $\mathbb{L}$. Plugging $k = 2$ into (6.16) yields again the generating function of the ladder also computed in Example 15.

Bibliography

[1] Alm, Sven Erick and Svante Janson: *Random self-avoiding walks on one-dimensional lattices.* Comm. Statist. Stochastic Models, 6(2):169–212, 1990, ISSN 0882-0287. https://doi.org/10.1080/15326349908807144.

[2] Beffara, Vincent and Cong Bang Huynh: *Trees of self-avoiding walks.* 2017. https://arxiv.org/abs/1711.05527v1.

[3] Ceccherini-Silberstein, Tullio and Wolfgang Woess: *Growth-sensitivity of context-free languages.* Theoret. Comput. Sci., 307(1):103–116, 2003, ISSN 0304-3975. https://doi.org/10.1016/S0304-3975(03)00095-1, Words.

[4] Chomsky, N. and M. P. Schützenberger: *The algebraic theory of context-free languages.* In *Computer programming and formal systems*, pages 118–161. North-Holland, Amsterdam, 1963.

[5] Dangovsk, Rumen and Chavdar Lalov Lalov: *Self-avoiding walks of lattice strips.* 2017. https://arxiv.org/abs/1709.09223v1.

[6] Dangovski, Rumen: *On self-avoiding walks on certain grids and the connective constant.* Serdica Math. J., 38(4):615–632, 2012, ISSN 1310-6600.

[7] Duminil-Copin, Hugo and Stanislav Smirnov: *The connective constant of the honeycomb lattice equals $\sqrt{2 + \sqrt{2}}$.* Ann. of Math. (2), 175(3):1653–1665, 2012, ISSN 0003-486X. https://doi.org/10.4007/annals.2012.175.3.14.

[8] Flajolet, Philippe and Robert Sedgewick: *Analytic combinatorics.* Cambridge University Press, Cambridge, 2009, ISBN 978-0-521-89806-5. https://doi.org/10.1017/CBO9780511801655.

[9] Flory, Paul J.: *Principles of Polymer Chemistry.* Cornell University Press, 1953, ISBN 0-8014-0134-8.

[10] Grimmett, Geoffrey: *Percolation.* Springer, New York, 1989, ISBN 0-387-96843-1.

[11] Grimmett, Geoffrey: *Three theorems in discrete random geometry.*
Probab. Surv., 8:403–441, 2011, ISSN 1549-5787. https://doi.org/
10.1214/11-PS185.

[12] Grimmett, Geoffrey and Zhongyang Li: *Self-avoiding walks and the
Fisher transformation.* Electron. J. Combin., 20(3):Paper 47, 14, 2013,
ISSN 1077-8926.

[13] Grimmett, Geoffrey R. and Zhongyang Li: *Locality of connective constants.* 2014. https://arxiv.org/abs/1412.0150.

[14] Grimmett, Geoffrey R. and Zhongyang Li: *Bounds on connective
constants of regular graphs.* Combinatorica, 35(3):279–294, 2015,
ISSN 0209-9683. https://doi.org/10.1007/s00493-014-3044-0.

[15] Grimmett, Geoffrey R. and Zhongyang Li: *Cubic graphs and the golden
mean.* 2016. https://arxiv.org/abs/1610.00107v2.

[16] Hammersley, J. M.: *Percolation processes. II. The connective constant.*
Proc. Cambridge Philos. Soc., 53:642–645, 1957.

[17] Jacobsen, Jesper Lykke, Christian R. Scullard, and Anthony J.
Guttmann: *On the growth constant for square-lattice self-avoiding
walks.* J. Phys. A, 49(49):494004, 18, 2016, ISSN 1751-8113. https:
//doi.org/10.1088/1751-8113/49/49/494004.

[18] Jensen, Iwan: *Improved lower bounds on the connective constants for
two-dimensional self-avoiding walks.* J. Phys. A, 37(48):11521–11529,
2004, ISSN 0305-4470. https://doi.org/10.1088/0305-4470/37/
48/001.

[19] Madras, Neal and Gordon Slade: *The self-avoiding walk.* Mod-
ern Birkhäuser Classics. Birkhäuser/Springer, New York, 2013,
ISBN 978-1-4614-6024-4; 978-1-4614-6025-1. https://doi.org/10.
1007/978-1-4614-6025-1, Reprint of the 1993 original.

[20] Nienhuis, Bernard: *Exact critical point and critical exponents of* $O(n)$
models in two dimensions. Phys. Rev. Lett., 49(15):1062–1065,
1982, ISSN 0031-9007. https://doi.org/10.1103/PhysRevLett.
49.1062.

[21] Pönitz, André and Peter Tittmann: *Improved upper bounds for self-
avoiding walks in* $\mathbf{Z}^d$. Electron. J. Combin., 7:Research Paper 21, 10,

2000, ISSN 1077-8926. http://www.combinatorics.org/Volume_7/Abstracts/v7i1r21.html.

[22] Zeilberger, Doron: *Self-avoiding walks, the language of science, and Fibonacci numbers.* J. Statist. Plann. Inference, 54(1):135–138, 1996, ISSN 0378-3758. https://doi.org/10.1016/0378-3758(95)00162-X.

Printed in the United States
By Bookmasters